Communication Strategies of Governments and NGOs

Manuel Adolphsen

Communication Strategies of Governments and NGOs

Engineering Global Discourse at High-Level International Summits

Springer VS

Manuel Adolphsen
Mannheim, Germany

Inaugural dissertation submitted to the School of Humanities at the University of Mannheim, 2012

OnlinePlus material for this book is available on the following link
http://www.springer-vs.de/978-3-658-05503-5

ISBN 978-3-658-05503-5 ISBN 978-3-658-05504-2 (eBook)
DOI 10.1007/978-3-658-05504-2

The Deutsche Nationalbibliothek lists this publication in the Deutsche Nationalbibliografie; detailed bibliographic data are available in the Internet at http://dnb.d-nb.de.

Library of Congress Control Number: 2014936409

Springer VS

Printed on acid-free paper

Springer VS is a brand of Springer DE.
Springer DE is part of Springer Science+Business Media.
www.springer-vs.de

Preface

This book presents the results of my PhD project, which I worked on while being a research associate at the University of Mannheim between 2008 and 2012. Living in Mannheim and going through the often lengthy and strenuous tasks associated with doing a PhD was not always easy. As a matter of fact and for a number of reasons, Mannheim was a rather restless and challenging chapter of my life so far.

But what is also true is that my time in Mannheim was tremendously valuable. Intellectually, it taught me to fully comprehend and appreciate the beauty and usefulness of scientific thinking – a true gift for life. It made me discover the fun of teaching and the satisfaction of developing new knowledge from scratch. And on a personal level, Mannheim meant the emergence of close friendships that I will never want to miss. There were countless warm, fun moments shared with wonderful people – and I am grateful for every single one of them.

First and foremost, I want to thank all my colleagues at the Institute of Media and Communication Studies for their longstanding companionship and inspiring ideas, including (in alphabetical order) Alice, Andreas, Anja, Antal, Bine, Eike, Franzi, Julia, Kathrin, Imme, Madeline, Maria, Marianne, Leonard, Nicole, Nils, Professor Kohring, Professor Vorderer, and our numerous fantastic student assistants who were always of great help. Special thanks go to Bine, Eike, and – of course – the world's best office mate Maria (with Evan) for valuable moments and a hell of a lot of fun. Professor Hartmut Wessler provided me with relaxed yet target-oriented supervision and was simply an inspiring boss and good friend. He (along with wonderful Marita) deserves special thanks, too.

And finally, there were numerous marvelous people beyond the microcosm of 'MKW' – some of them in Mannheim, many of them in Berlin, most of them scattered around the globe – who supported me in times void of motivation and granted me those moments of joyful distraction and authentic friendship required for pulling off a PhD. Christian, Dominik, Elias, Fred, Jenny, Julia, Lukas, Micha, and Sebastian (with Sarah) are certainly to be mentioned here, but there were many, many more – including my caring parents Catha and Andres.

Berlin, December 2013 Manuel Adolphsen

Table of contents

List of tables		11
List of figures		13
List of abbreviations		15
1.	**Summits in an age of global governance**	17
1.1	Research focus and normative implications	21
1.2	Research questions	22
1.3	Overall assumptions and clarification of terms	23
1.4	Structure of the study	25
2.	**Public discourse beyond national borders**	27
2.1	Public discourse and the notion of the public sphere	27
2.1.1	Modeling the public sphere as network of fora	31
2.1.2	The public sphere and the national context	37
2.2	Different perspectives on transnational public spheres	38
2.2.1	Transnational public spheres as autonomous 'superstructures'	39
2.2.2	Transnational public spheres as outcome of transnationalization	44
2.3	Transnationalization of national public spheres	46
2.3.1	Long-term restructuring of discursive communication	47
2.3.1.1	Operationalizing transnationalization	48
2.3.1.2	Transnationalization around the issue of climate change	50
2.3.2	Short-term episodes of ritualistic communication	52
2.4	The summit as transnational forum and provider of symbolic resources	57
3.	**Summits as strategic communication resources**	59
3.1	Public diplomacy: strategic communication on the international stage	59
3.1.1	Definitional issues	60
3.1.2	Previous research: governments' public diplomacy	62

3.1.3	Previous research: NGOs' public diplomacy	66
3.2	Conceptualizing high-level international political summits (HIPS)	71
3.2.1	A closer look: History and functions of summits	74
3.2.2	Provision of symbolic resources	75
3.2.3	Participation of civil society	78
3.3	Public diplomacy at HIPS	79
3.3.1	Previous research	79
3.3.2	Conceptual model of the development of communication strategies	86
3.4	Summary of theoretical argument	92
4.	**Investigating strategic communication at summits**	95
4.1	The case study approach: basic features and specific configuration	95
4.1.1	Causality in case study research	96
4.1.2	Properties and value of case study findings	97
4.1.3	Nomenclature: 'classes of events', 'units', and 'cases'	98
4.1.4	Integrating process-tracing and structured, focused comparison	99
4.2	Unit selection: the Cancún summit and the climate change debate	101
4.3	Case selection: governments and NGOs at the Cancún summit	103
4.4	Data collection	OnlinePLUS
4.4.1	Core method: semi-standardized interviews	
4.4.1.1	Development of interview guides	
4.4.1.2	Recruitment of interviewees	
4.4.1.3	Sequence, mode, and setting of interviews	
4.4.1.4	Securing confidentiality	
4.4.2	Supplementary data: on-site observation and collection of PR material	
4.5	Data analysis	OnlinePLUS
4.5.1	Transcription	
4.5.2	Qualitative content analysis	
4.5.2.1	Computer-aided qualitative content analysis	
4.5.2.2	Steps of analysis	
4.6	Chapter summary	105

5.	**Structuring the analysis and laying foundations**	107
5.1	'Dissecting' the summit: the structure of the analysis	107
5.2	Setting and infrastructure of the 2010 UN climate summit	109
5.2.1	Organization and participation	109
5.2.2	Location and venues	111
5.2.3	Communication infrastructure	113
5.3	Chapter summary	116
6.	**Political actors' communication structures, processes, and preplanned strategies**	117
6.1	Political actors' communication structures and processes	OnlinePLUS
6.1.1	Government delegations	
6.1.1.1	Brazil	
6.1.1.2	Germany	
6.1.1.3	India	
6.1.1.4	Mexico	
6.1.1.5	South Africa	
6.1.1.6	United States	
6.1.2	NGO delegations	
6.1.2.1	Climate Action Network (CAN) International	
6.1.2.2	Friends of the Earth International (FOEI)	
6.1.2.3	Global Campaign for Climate Action (GCCA)	
6.1.2.4	Greenpeace International	
6.1.2.5	World Wide Fund for Nature (WWF) International	
6.2	Political actors' communication strategies and activities	117
6.2.1	Governments' messaging strategies	117
6.2.2	NGOs' messaging strategies	124
6.2.3	A closer look: Dissent over strategy among NGOs	128
6.2.4	Media relations activities	130
6.2.5	Public outreach activities	138
6.2.6	A closer look: NGOs' staging of photo opportunities	146
6.3	The influence of long-term factors on strategy choice	151
6.3.1	Type of actor	152
6.3.2	Worldviews and standpoints	153
6.3.3	Degree of professionalization	155
6.3.4	Individual background of PR professionals	157
6.4	A typology of summit PR approaches	158
6.5	Chapter summary	160

7.	**Short-term interactions between summit PR and summit proceedings**	163
7.1	Short-term effects of political actors' PR at the summit	163
7.1.1	The relationship between PR professionals and journalists	164
7.1.1.1	NGO communicators as experts and information brokers	164
7.1.1.2	Collaborative composing of visual coverage	167
7.1.1.3	Informal interactions in shared workspace	167
7.1.2	The influence of summit coverage on summit proceedings	168
7.1.2.1	Media attention as constitutive element of the summit	169
7.1.2.2	Media setting the tone of the summit	169
7.1.2.3	Media as negotiation tools	172
7.2	Short-term discursive opportunity structures	175
7.3	Political actors' learning effects	186
7.4	Chapter summary	191
8.	**The summit as world stage?**	193
8.1	Summary of empirical findings	193
8.2	Normative assessment of empirical findings	203
8.3	Concluding theoretical remarks	210
Literature		213
Appendices		OnlinePLUS
I	Recruiting email	
II	Supporting letter	
III	Interview guide t^1	
IV	Interview guide t^2	
V	Interview guide t^3	
VI	Interview guide UNFCCC	
VII	Transcription rules	
VIII	Coding frame	

All chapters marked 'OnlinePLUS' are not included in this book but can be downloaded from this book's companion website at www.springer.com.

List of tables

Table 1.1 Clarification of terms related to transnational communication 24

Table 3.1 Defining features and optional features of high-level international summits 73

Table 6.1 Recipients of CAN International's Fossil of the Day award 144

Table 8.1 Summary of normative assessment of empirical findings 209

List of figures

Figure 2.1	Contrasting theoretical perspectives on transnational public spheres	40
Figure 2.2	Theoretical approaches to the transnationalization of public spheres	47
Figure 3.1	Three variants of public diplomacy	65
Figure 3.2	The 'boomerang pattern' in NGO public diplomacy	70
Figure 3.3	Routes of strategic communication at summits	86
Figure 3.4	Deductively developed conceptual model (prior to empirical testing)	89
Figure 4.1	Methodological nomenclature used in this study	100
Figure 4.2	Timeline listing all interviews conducted	OnlinePLUS
Figure 4.3	Screenshot of MAXQDA software	OnlinePLUS
Figure 4.4	Steps of qualitative content analysis	OnlinePLUS
Figure 5.1	Links between chapter sections and components of conceptual model	108
Figure 6.1	Spectrum of governments' messaging strategies	118
Figure 6.2	Spectrum of NGOs' messaging strategies	129
Figure 6.3	Brazilian country pavilion, *Espaço Brasil*	140
Figure 6.4	Summit website run by Mexican host government	140
Figure 6.5	US country pavilion, U.S. Center 2010 Cancun	141
Figure 6.6	German information booth	142
Figure 6.7	Hot-air balloon photo op by Greenpeace International	147
Figure 6.8	Pyramid of Hope photo op by GCCA	148
Figure 6.9	Life-ring photo op by Greenpeace International and GCCA	149
Figure 6.10	Typology of governments' summit PR approaches	159
Figure 6.11	Typology of NGOs' summit PR approaches	160

Figure 7.1 Different avenues for media involvement in the communication between two summit actors 171
Figure 7.2 Fake contact ad in December 1, 2010 issue of the *ECO* newsletter 179
Figure 7.3 'I ♥ KP' photo op by GCCA 180
Figure 7.4 Protest ad in December 10, 2010 issue of the *Financial Times* 182
Figure 7.5 Protest ad in December 14, 2007 issue of the *Jakarta Post* 184

Figure 8.1 Revised conceptual model (after empirical testing) 202

List of abbreviations

9/11	September 11, 2001 terrorist attacks in the US
AFP	Agence France-Presse (French news agency)
AOSIS	Alliance of Small Island States
AP	Associated Press (US news agency)
APTN	Associated Press Television News
ASEAN	Association of Southeast Asian Nations
AU	African Union
BASIC	Alliance between Brazil, India, China, and South Africa
BBC	British Broadcasting Corporation
BRICS	Alliance between Brazil, Russia, India, China, and South Africa
CAN	Climate Action Network
CDM	Clean Development Mechanism
CDN	Companhia de Noticias (Brazilian PR firm)
CNN	Cable News Network
CNNI	CNN International
COP	Conference of the Parties (annual meeting of the member states of the UNFCCC, also referred to as 'UN Climate Change Conference' or 'climate summit')
COP-1	1995 climate summit in Berlin, Germany
COP-13	2007 climate summit in Bali, Indonesia
COP-15	2009 climate summit in Copenhagen, Denmark
COP-16	2010 climate summit in Cancún, Mexico
COP-17	2011 climate summit in Durban, South Africa
EBU	European Broadcasting Union
ECCP	European Climate Change Programme
EFE	Agencia EFE (Spanish news agency)
e.g.	*exempli gratia*, 'for example'
ENGO	Environmental non-governmental organization
EU	European Union
ENG	Electronic news gathering (technical term used in the TV industry)
FIFA	Fédération Internationale de Football Association
FOE	Friends of the Earth

GCCA	Global Campaign for Climate Action
HIPS	High-level international political summits
HRW	Human Rights Watch
ibid.	*ibidem* (referring to the same source referenced in the preceding citation)
i.e.	*id est*, 'that is'
IBC	International Broadcast Center (temporary working facility for broadcasters during large-scale events, such as COPs)
IGO	Intergovernmental organization
IMF	International Monetary Fund
IPCC	Intergovernmental Panel on Climate Change
IPS	Inter Press Service ('alternative' news agency focusing on issues of the Global South and globalization)
Mercosur	Common Southern Market (economic and political agreement between Argentina, Brazil, Paraguay, and Uruguay)
MPH	Make Poverty History campaign
NAFTA	North American Free Trade Agreement
NGO	Non-governmental organization
PCG	Political coordination group (body within CAN International)
PR	Public relations
QDA	Qualitative data analysis (usually referring to computer-aided forms)
SECOM	Secretariat of Social Communication (Brazilian government communication agency)
TV	Television
UK	United Kingdom of Great Britain and Northern Ireland
UN	United Nations
UNFCCC	United Nations Framework Convention on Climate Change
UNICEF	United Nations Children's Fund
US	United States of America
USSR	Union of Soviet Socialist Republics
VoIP	Voice over Internet Protocol (also known as 'Internet telephony')
vs.	versus
WEF	World Economic Forum
WTO	World Trade Organization
WWF	World Wide Fund for Nature

1. Summits in an age of global governance

The international system has witnessed fundamental change over recent decades. Since the end of the Cold War and its bloc-based confrontation, a more complex constellation of multipolarity has emerged, featuring the rise of several ambitious countries eager to make an impression on the world stage. The influence of 20th-century hegemons has simultaneously been weakened (or at least transformed in nature), turning the coordination of state interests for purposes of international policy-making into an intricate and often challenging affair. Significant parts of such processes have also been moved to dedicated institutions at the supranational level, such as IGOs within the UN framework or regional alliances like ASEAN, AU, or EU. Despite being mandated by state principals, such agents might pursue independent agendas and constitute autonomous political voices in the international system.[1] A third type of actor to be met in this domain is large NGOs like Human Rights Watch or Greenpeace International that typically represent moral or progressive concerns and aim at inserting them into international policy debates.

Scholars like Rosenau (1995) or Held and McGrew (2002) use the term 'global governance' for describing the international political environment in the new millennium. In their perspective, the traditional, primarily state-based international relations of the Cold War have given way to a new fragile system of actors situated on various levels and equipped with varying degrees of authority and legitimacy. The common characteristic binding states, IGOs, and NGOs together is their stake in influencing decisions of international relevance; instead of a set of like units (a basic assumption of realist thought in international relations theory), contemporary global governance consists of a colorful "nexus of systems of rule-making, political coordination and problem-solving which transcend states and societies" (Held and McGrew, 2002, p. 8), a "patchwork of overlapping jurisdictions" (p. 10) that generates "ambiguities about the principal location of authority and political responsibility" (ibid.).

The politics of global governance deal with a set of deterritorialized issues that did not play a large role in traditional international relations. Cold War for-

1 This refers to the classic principal-agent model in political science. See Hix (2005, pp. 27-31) for an application to the behavior of supranational institutions.

eign politics dealt with matters of confrontation and strategic influence; global governance, in contrast, is about coordination, cooperation, and problem-solving. Environmental pollution, human rights, drug trafficking, and terrorism are examples of the new political substance "of transnational policy issues which cut across territorial jurisdictions and existing political alignments, and which require international cooperation for their effective resolution" (p. 6). According to Weiss (2008), the international scope of these issues, whose handling lies far beyond what individual states can achieve on their own, is the real driver of global governance. For him, the concept simply comprises all "collective efforts to identify, understand, or address worldwide problems that go beyond the capacity of individual states to solve" (p. 219). While this definition can surely cover a wide range of multi-level activity by all sorts of organizations, it focuses our attention on the prevalence of selected issues of global relevance that require concerted efforts of political actors around the globe.

Global Governance around the issue of climate change

One of such global issues is climate change. Carbon dioxide emissions, commonly regarded as key factor in anthropogenic manipulation of atmospheric conditions, cannot be contained by national borders; their long-term impact may be experienced around the world, regardless of where they were generated. Essentially, climate change is cumulatively caused by the whole world and can hence only be solved by the whole world (Beisheim, 2004, pp. 101-102). At the same time, the particular response that humankind should give in mitigating and adapting to the threat of climate change is far from clear. The global nature of the challenge does not automatically translate into a common global agenda; international climate policy is instead driven by vastly different conceptions of the seriousness of the threat, the ideal course of action, and the attribution of responsibility. This has to do with the fact that, despite their inherently global nature, processes of climate change may have differing implications in different parts of the world (Yearley, 1995, p. 227). "Because of climate, altitude and other geographical factors, their impacts will be greater in some areas than in others. Furthermore, on average, wealthier societies and the wealthier people in societies will be better placed to withstand their impacts than will other groups" (ibid.).

This has turned climate change into a hotly debated issue. A multifaceted system of political groupings and authorities at the transnational level accommodates diverse political processes that aim at finding a policy response to climate change. "Transnational climate governance" (Andonova, Betsill, and Bulkeley, 2009, p. 52) has gradually emerged since the 1960s and is nowadays primarily built on the United Nations Framework Convention on Climate Change (UN-

FCCC) and a plethora of associates and stakeholders, e.g., the Intergovernmental Panel on Climate Change (IPCC), regional initiatives like the European Climate Change Programme (ECCP), or diverse nongovernmental groups with an interest in shaping global climate policy (Beisheim, 2004, pp. 48-53). Sovereign governments represent one particular type of actor to be found in this specific network of global governance; they are continuously exposed to the influence of players that have started to operate outside national logics. Yearley (1995) argues that the rising significance of "such supranational action, whether by groups of government, by consortia of influential campaign bodies or by industrial groups and their lobbyists, indicates that in a globalizing world, political initiative and political control can begin to pass away from the nation-state and into the hands of supranational actors" (p. 212).

A lack of legitimacy in global governance

Critical observers of political decision-making at the supranational level have often raised questions regarding the democratic character of these processes. In democratic nation-states, institutions like elections and referenda with binding results, political parties aggregating and representing citizens' interests, or free media exposing power holders' misdemeanors constitute a basic prerequisite for the correspondence of the making and implementation of policy with the self-determined will of the majority of citizens. Scholars of democracy regard such conditions as facilitators of 'input legitimacy', the fundamental accountability of those who govern to the governed and a rootedness of the former's decisions in the latter's preferences (see, e.g., Schmidt, 2010, p. 15 for details).

With regard to transnational politics, however, it has often been argued that input legitimacy is limited (see, e.g., Scharpf, 1998). Legally institutionalized procedures that could make supranational policy-making responsive and accountable to citizen preferences, such as the inclusion of national parliamentarians or implementation of supranational plebiscites, do not exist in global governance (Imber, 1997, pp. 226-228). Also, the argument that governments active on the supranational level can be seen to act on behalf of their peoples (provided they carry a democratic mandate) is contestable, as citizens or their parliamentarians are not "briefed by ministers departing for Brussels or New York, revealing in advance, and in public, the limits to which the government will go to secure agreement in an EU or UN forum" (ibid., p. 221). As laid out in realist thought, governments negotiating with other governments adhere to other principles than full representation of their citizens (McGrew, 1997, pp. 16-17). Such realities of global governance lead Habermas (2001a) to conclude that the "new forms of international cooperation lack the degree of legitimation even remotely approaching the requirements for procedures institutionalized via nation-states" (p. 71). In short, the 'postnational constellation' of today's world features serious "legitimation gaps" (ibid.).

Summits as generators of legitimacy in global governance?

One stream of discussion around how the democratic deficit in global governance could be offset centers on notions of public discourse and the public sphere (both concepts are investigated in detail at a later point). In this perspective, transnational politics should be complemented by processes of public debate, in which political actors present their standpoints and proposals and make them available for discussion by other actors or citizens. As public exchanges of positions among actors, as procedures of making a case, commenting, and revising, such processes might generate the type of legitimacy that is not provided by institutional routes within global governance. Accordingly, various authors emphasize that "critical transnational debate on globalization" (Beyeler and Kriesi, 2005, p. 107) is needed, that "the emergence of public debates that critically observe […] and hold accountable the decision makers on the transnational and supranational level seems of great normative importance" (Wessler, 2010, ¶ 61). The nature and normative quality of such debate is also discussed in the context of what is referred to as international, transnational, or even global 'public sphere', a nonphysical space of communication around politically relevant issues that spans across national borders or even the entire globe. The legitimacy of global governance, in this view, is reliant on the existence of such spaces; if they are absent, transnational politics take place in a democratic void: "Without a flourishing of international public sphere, the global sociopolitical order becomes defined by the realpolitik of nation-states that cling to the illusion of sovereignty despite the realities wrought by globalization" (Castells, 2008, p.80).

This research highlights one specific way in which transnational public spheres may emerge. It explores high-level international political summits (HIPS), particularly those conducted yearly within the UNFCCC framework, as catalysts of temporary, issue-bound public spheres of transnational nature. The fundamental assumption, which is backed up theoretically at a later point, is that their worldwide relevance and elevated symbolic meaning turn these 'climate summits' into events that are simultaneously observed and discussed worldwide. They can be seen as are rare occasions at which regimes and issues of global governance are indeed accompanied by public discourse, at which public debate across national borders, which is deemed so important for the legitimation of transnational politics, might actually emerge. In this perspective, HIPS and their triggering of short-term, issue-specific transnational public spheres might carry democratizing potential for global governance.

Public discourse, in the understanding of this study, commonly has a strategic dimension to it. Actors who contribute to the debate, who choose to introduce their position into this exchange of positions usually aim at attaining particular objectives – for example, the raising of particular issues or the building of support for specific initiatives. Contributions are hence preceded by internal preparations, by reflections on public discourse's desired consequences. Such strategizing is faced with specific opportunities and challenges if public discourse is of transnational nature, for example, as part of HIPS. Above all, these events can be regarded as resources for political actors; by engaging in transnational public discourse triggered by HIPS, they efficiently publicize their positions across national borders and thereby attain a geographical spread of their messages that could not be easily achieved by other means. On the other hand, the influencing of transnational public discourse around HIPS might also call for specific approaches and techniques that are not required on the national level.

1.1 Research focus and normative implications

Generally, the current state of knowledge on strategic communication at HIPS is limited and requires further investigation. In particular, the way in which political actors make use of these events and their cross-border reach for specific communication strategies has been neglected in prior research. While the significant role of high-visibility events in actors' international PR has been hinted at by some scholars (e.g., Gilboa, 2001), the exact nature of attempts to 'engineer' transnational public discourse at these occasions remains unclear. One particular study with a related interest and design is Betsill and Corell's (2008a) investigation of the influence of accredited NGOs at environmental summits. However, instead of the shaping of public discourse, they focus on direct, i.e. 'invisible', influences on negotiators. Publicly visibly NGO campaigning at summits is analyzed in a small number of other studies, but either without an explicit view to transnational public discourse (e.g., Sireau, 2009) or based on methodologies focusing on the resonance of summit PR with the media, instead of circumstances of production (e.g., Beyeler and Kriesi, 2005). Governmental communication at these events seems to be ignored altogether.

This study aims at filling this research gap. It investigates the shaping and implementation of governments' and NGOs' communication strategies at HIPS, based on a case study of the 2010 UN climate summit in Cancún, Mexico. It aims at identifying and exploring the factors that determine how these actors exploit the summit as transnational communication resource. Employing such focus, this research is at first sight rooted in a non-normative and analytical perspective with some rational-choice influences: Actors are assumed to rationally reflect on available options for developing strategies that are most conducive to attaining preset goals (see Esser, 1999, pp. 296-298). However, I do not abandon the normative contextualization of HIPS as potential generators of legitimacy in global governance, which was introduced above. In the conclusion (see section 8.2), I scrutinize the normative qualities of HIPS as resources for transnational political PR: Do the kind of public communication originating at HIPS and political actors' attempts to influence it really justify the attribution of a legitimacy-creating role to HIPS? How can strategic communication at HIPS be evaluated normatively? Hence, in an innovative way, this study integrates a normative perspective inspired by democratic theory with a positive, analytical approach mildly influenced by rational-choice thinking.

1.2 Research questions

In line with the general research focus, the specific research questions are:

- How do political actors of state and non-state nature carry out strategic communication at high-level international political summits (HIPS), such as the 2010 UN climate summit in Cancún, Mexico? How do their explicit communication strategies and corresponding communication activities look like? How do they consider the event's transnational reach in their strategies? To what extent are strategies made subject to ad-hoc modification during the course of the summit?
- Why are particular communication strategies chosen? Which factors shape the development of communication strategies? Why are communication strategies made subject to ad-hoc modification during the course of the summit (if at all)?

Different purposes of research loom behind the two sets of questions. The questions of the first set ('how?', 'to what extent?') are descriptive in nature and hence oriented towards the empirical exploration and structured presentation of phenomena of social reality. In contrast, the second set of questions ('why?', 'which factors?') reveals explanatory ambitions and is aimed at the identification of factors bringing about particular manifestation of phenomena of social reality. Explanatory research commonly builds on descriptive research: Without a thorough exploration and description of reality, it is hard to establish explanations.

1.3 Overall assumptions and clarification of terms

A variety of concepts and terms has been introduced thus far (for example, public discourse, public sphere, HIPS, NGO, PR etc.). Most of them are discussed in detail throughout the theory chapters. However, a number of basic assumptions and clarifications should be introduced at this point:

When I speak of political actors, I refer to collective actors that are involved – or aim at being involved – in the shaping of collectively binding decisions. In particular, I focus on national governments and international NGOs and their respective delegations at HIPS. While I acknowledge that these are highly complex and differentiated organizations characterized by elaborate hierarchies and internal friction, I treat them as unitary actors that are, for the most part, oriented towards attaining particular organizational objectives. My investigation of such strategic organizational objectives is based on observations of or information provided by organizations' individual members, usually PR professionals. Generally, I assume their actions and motives to be reflective of the entire organization's outlook, although I do acknowledge that individual members might act outside the parameters set by organizational goals. However, in the case of PR professionals, whose job is to speak on behalf of the entire organization, this can be considered more unlikely.

Secondly, strategic communication refers to acts of communication that an actor initiates in pursuit of objectives, "the purposeful use of communication by an organization to fulfill its mission" (Hallahan, Holtzhausen, van Ruler, Verčič, and Sriramesh, 2007, p. 3). The fundamental motive behind engaging in strategic communication is to exercise influence on social circumstances. Strategic communication is considered effective (or successful) if such consequences can indeed by detected. Generally, strategic communication refers to a basic orientation of communication; it is not limited to particular forms or instruments, such as advertising or media relations (p. 10). Instead, it may draw on a multitude of modes and ways in which communication can be conducted. Depending on the objective, strategic communication may even comprise consensus-oriented forms of interaction, such as roundtables or mediation efforts, even though in classic Habermasian thought, they would be counted as "pseudoconsensual" (1979, p. 210). Specific configurations of the various parameters of communication (such as message, channel, or targeted audience) for strategic purposes are here called communication strategies; they involve elements of (1) a priori planning, which takes place inside organizations and is thus mostly hidden to outsiders, and (2) public implementation, which commonly crosses the boundaries to the organization's external environment and is visible to outsiders. Together, organizational practices of planning and implementation form strategic communication processes.

Finally, the attribute 'transnational' refers to social processes that cross national borders without being significantly influenced by this crossing. Hence, these processes take place beyond or 'above' national borders, with the latter possessing only little structuring power. 'International', in contrast, refers to exchanges between countries; here, national borders are an important shaper of social processes (Wessler and Brüggemann, 2012, p. 3). Thus, if we talk about international communication, this refers to exchanges among actors rooted in different countries. The actors of transnational communication, however, while also operating from particular countries, cannot be viewed as genuinely national, as the national context does not strongly affect their communication (Thörn, 2007, p. 899); transnational communication hence occurs outside of national borders and contexts.

Level of communication	**Scope of communication**	**Equivalent in governance**
National	Bounded by country	National
International, i.e. between countries – national borders possess structuring power for communication	Bounded by exchanges between two or more countries	Intergovernmental – powers are with national governments, which use them in accordance with treaties/institutions formed with other national governments
Transnational, i.e. beyond/'above' countries – national borders lack structuring power for communication	Bounded by particular regions, issues etc. beyond countries – potentially even global*, i.e. evenly spanning all continents	Supranational – selected powers are delegated to institutions on a higher level, which in turn set decisions to be obeyed by national governments

*in some less precise accounts, 'transnational' appears to be hastily equated with 'global'

Table 1.1: Clarification of terms related to transnational communication

While 'international' and 'transnational' hence refer to different levels or qualities of communication, it is important to distinguish them from scope or "geographical extent" (Wessler, Peters, Brüggemann, Kleinen-von Königslow & Sifft, 2008, p. 23). This is often confused in pertaining literature. While 'transnational communication' might suggest a relation to a larger number of countries around the world, such connotation should be treated with caution. Transnational communication is not equal to global communication; neither does it automatically involve more countries than international communication. Level and scope of communication are distinct dimensions of communication that should be kept apart conceptually (see table 1.1).

As described earlier, it is assumed here that HIPS like the UN climate summits may trigger public discourse of transnational nature and may hence be exploited for transnational strategic communication. While surely not all actors at these events are genuinely transnational – as a matter of fact, the overwhelming majority of participants is constituted by national governments and national media –, the global relevance and 'denationalized' setting of these events may facilitate communication processes that are not primarily related to specific countries. Ultimately, this is an empirical question that is to be explored in this research.

More background on these notions and terms is provided in the theoretical part (chapters 2 and 3), which also presents definitions for most of the other concepts raised so far, including public discourse (see p. 28), public sphere (p. 31), PR (p. 33), public diplomacy (p. 62), HIPS (p. 72), and NGO (p. 78).

1.4 Structure of the study

The structure of this study is generally oriented towards the archetypal research process in a deductive, Popperian framework. The identification of research questions requiring investigation in the introduction is followed by the exploration of applicable theories (chapters 2 and 3). The theoretical part is structured along three theses, which focus the examination of prior research, and finishes with a conceptual model of how political actors' development of summit communication strategies could be explained. This model takes the place of what in other studies are the hypotheses; it guides my empirical work and posits relations of influence that are to be verified. Subsequently, details regarding the design and methods of this research are clarified (chapter 4). The findings are discussed in three empirical chapters, which depict the infrastructural context of the Cancún summit (chapter 5), investigate more structural, preset aspects of strategic summit communication (chapter 6), and reconstruct more procedural, dynamic interactions between summit actors (chapter 7). Finally, the major empirical findings are summarized and incorporated in a revised version of the conceptual model. The findings are also assessed against a normative background and supplemented by some concluding theoretical remarks (chapter 8).

2. Public discourse beyond national borders

This and the following chapter introduce the theoretical context for this study. They integrate several strands of research into one coherent argument, which is built in three steps:

I start by introducing the notions of public discourse and public sphere as the context of strategic communication efforts (2.1). Subsequently, I explore how these notions can be conceived of at the transnational level (2.2 and 2.3). This leads me to the interim conclusion that *HIPS constitute short-term fora that interconnect with and temporarily transnationalize national public spheres* (2.4).

In the subsequent chapter, I explore previous research on strategic communication beyond national borders – also known as public diplomacy – (3.1), on political summits (3.2), and on how public diplomacy is conducted at summits (3.3). This is to show that *from a strategic communication perspective, the transnationalizing potential of HIPS turns them into valuable resources for political actors' public diplomacy efforts.*

Looking into various factors influencing public diplomacy in that context, I finally argue that *actors' choice of communication strategies at HIPS depends on both structural/long-term and dynamic/short-term factors* (3.3.2). This is summarized in a conceptual model taking the place of what constitutes the hypotheses in other studies. The purpose of subsequent chapters is then to empirically test and revise this model.

Both chapters are tailored towards rooting all three theses in relevant research. They are hence written with argumentative intent and feature an appropriate selection of literature.

2.1 Public discourse and the notion of the public sphere

As mentioned above, in the eyes of some scholars, the work of global governance regimes, like the one around the issue of climate change, should be accompanied by public discourse across national borders. Only through such processes can decisions of global relevance be met with appropriate public scrutiny around the world and supranational institutions, such as the UNFCCC, obtain democratic legitimacy. In basic non-normative understanding, public discourse simply refers to

speakers' exchange of speech acts in front of an audience.

These speech acts relate to each other and are made with regard to issues of wider relevance, which are often of conflictual nature (Ferree, Gamson, Gerhards, and Rucht, 2002, p. 9; Raupp, 2011, p. 101). This non-normative definition does not make any claims about a socially desirable quality of public discourse. Such normative proposals come from, e.g., Habermas (1981, pp. 25-44) or Peters (2007, pp. 62, 89-97), who view public discourse as a mode of attaining mutual understanding through rational exchange of arguments. For them, it is fundamentally different from, for example, targeted bargaining within negotiations and 'aesthetic expressions' in popular culture (Peters, 2007, p. 90) or strategic action shaped by "egocentric calculations of success" (Habermas, 1985, p. 286) taking the place of "acts of reaching understanding" (ibid.). Such normative conceptions are not in the focus of this study.

The public sphere as locus of public discourse

Closely linked to the notion of public discourse is that of the 'public sphere'. While the former emphasizes processes of communication on particular issues, the latter denotes the totality of these processes as well as the non-physical space in which they occur. The concept constitutes a key category in communication studies, and an analysis of actors' efforts to 'engineer' public discourse is not complete without its consideration. While the term appears in a broad variety of research,[2] several proposals for how the public sphere could be theorized share a conceptual core. As emphasized by the spatial connotation of the term 'sphere', it is about a non-physical space for society's debating of issues – "a field of communication" (Peters, Sifft, Wimmel, Brüggemann, and Kleinen-von Königslow, 2005, p. 140), in which speakers' exchange becomes accessible to audiences and public opinion is formed (Gerhards and Neidhardt, 1991, p. 42).

Inherent in this broad understanding is the assumption that there are certain issues which hold relevance to more than a few individuals. These issues concern the way in which people live *together* and therefore relate to the making of collectively binding decisions, i.e. politics (Gerhards, 1998, p. 268; Peters, 2007, p. 56). However, the standard for what is relevant to a wider public is flexible. A

2 Overviews can be found in, for example, Wimmer, 2007 or, more concisely, Donges and Imhof, 2005; a compilation of key texts in public sphere theory is provided by Gripsrud, Moe, Molander, and Murdock, 2010.

publicly debated issue can be of rather private nature, as scandals surrounding politicians' private conduct remind us. Hence, what is public about public issues is not so much their inherently 'serious' nature, but rather the fact that public speakers attribute to them some political relevance and therefore introduce them into the public sphere (Brüggemann, 2008, p. 40).

Also, in modern mass societies, the accessibility of such debates can only be sustained with the help of technical means. This is the purpose of the mass media (Ferree et al., 2002, p. 10; Gerhards and Neidhardt, 1991, p. 54; Kleinsteuber, 2000, p. 44), which through their bridging of distance, can "knit spatially dispersed interlocutors into a public" (Fraser, 2007, p. 10). Seen through the lens of public discourse, mass media provide the stage on which speakers can engage with each other. They turn discourse into *public* discourse and provide for the special quality of 'public-ness' that is not achievable otherwise (at least not on a wider scale). Due to this function, mass media are an integral feature of most parts of public sphere theory.

Different theoretical conceptions of the public sphere

Beyond these two assumptions, conceptions of the public sphere vary. A dividing line runs between *analytical* models suggesting how the concept could be approached for empirical study and *normative* accounts prescribing how the public sphere should function and positing particular qualities, for example for public discourse (see above). Generally, the notion of the public sphere grew out of Enlightenment thinking and has carried normative connotations ever since, even in conceptions lacking explicit normative claims (Gerhards and Neidhardt, 1991, p. 32). Its career in academia is inseparably linked with the name of German theorist Jürgen Habermas, who developed its understanding as a space for citizens' debate on communal problems, traditionally at such physical locations as bourgeois 18th-century coffeehouses (Habermas, 1990, pp. 90-107). Key to the early Habermasian conception is a focus on individual citizens consciously engaging in a particular quality of public discourse (here, the term *does* carry normative meaning) and essentially advancing society's welfare (Donges and Imhof, 2005, p. 158). The exchange of well-reasoned arguments brings about well-reasoned public opinion as foundation for political decisions (Gerhards, 1998, p. 268). Although analytical and normative elements clearly confound in this direct democratic conception (Gerhards and Neidhardt, 1991, p. 32), Habermas's work has remained influential as normative guidepost and point of reference in subsequent theorizing.

Contrasting normative conceptions focusing on individual agency and discourse, models inspired by systems theory regard the public sphere as particular environment within society, enabling its self-observation (Luhmann, 1990; Marcinkowski, 1993). Here, the public sphere should be seen less "within the context of the rise of the bourgeoisie than within that of the long-term changes in the general structure of society" (Gestrich, 2006, p. 428). Functional differentiation within society has brought about a set of subsystems running specialized internal discourses (ibid.); the function of the public sphere is to 'mirror' these systems and allow for the observation of their interaction. Observers may rely on the public sphere for identifying and possibly readjusting their position (Donges and Imhof, 2005, p. 156).

Gerhards and Neidhardt's (1991) conception of the public sphere as an intermediary system borrows from discourse models inspired by Habermas and those based on systems theory. This becomes clear from the dual approach of modeling the public sphere as system located between the political and other societal systems and also as the locus of political processes:

> "The public sphere is – on the occasion of elections, but also in between – a place of articulation for citizens' and interest groups' issues and opinions. It is a system contributing to the definition of the political system's agenda. Issues are set and respective opinions formed, which then suggest a direction for the political processing of these issues" (p. 40, own translation).

According to this conception, the public sphere is the place where issues and opinions are taken up (*input*), aggregated and structured into public opinion (*throughput*), and the latter passed on to the political system (*output*) (p. 35). It is intrinsically tied to the political system, constituting the breeding ground for democratically desirable values like transparency, popular control, or legitimacy (p. 41). If the public sphere fails in carrying out these functions, social movements may form in response (p. 80; Neidhardt, 1994, p. 8).

Besides this normative macro function, Gerhard and Neidhardt's design entails a view of the public sphere as network of fora. This notion goes back to Hilgartner and Bosk's (1988) analysis of "particular public arenas in which social problems are framed and grow" (p. 58) and has been published as 'arena model' (Neidhardt, 1994; Ferree et al., 2002) – yet without the normative connotation built into the conception as intermediary system. The model serves as conceptual cornerstone in this study, as it allows for insightful theorizing of the transnationalizing effects of summits. I hence pursue a theoretical path in this study that is neither embedded in normative conceptions of public discourse in the tradition of Habermas nor fully in line with perspectives stemming from systems theory. It is a positive, analytical perspective on the public sphere as "specific configuration of social communication" (Schäfer, Ivanova, and Schmidt, 2011, p. 134, own translation) that borrows from both theoretical strands. Furthermore, it does neither reduce the public sphere to a mere *collective* of speakers and audiences taking part in public discourse, nor does it only stand for the *state* of being public, or visible within a communicative space (Brüggemann, 2008, pp. 41-42).

2.1.1 Modeling the public sphere as network of fora

Essentially, Feree et al.'s (2002) arena model rests on a definition of the public sphere as a

network constituted by public fora and the communicative flows within and between them.

Such fora may differ in popular reach, topical scope, fixation of roles, or incorporation of distribution means (i.e., media). Through their linkages, they form a web of "interconnected arenas of public communication" (Brüggemann and Schulz-Forberg, 2009, p. 694), whose totality represents the public sphere. Conceptual tension may be found between the view of 'the' public sphere as a unitary, integrated space of communication and the notion of a more fragmented, networked environment featuring more central and more peripheral components (Brüggemann, 2008, p. 44; Latzer and Saurwein, 2006, p. 11). However, in a theoretical ideal, which was also laid out in later writings by Habermas (1992, p. 436), communication flows within this complex network are constantly synthesized and filtered, amplified and reacted to, supplemented and countered, so that, ultimately, *one* more or less integrated web of communication emerges: *the* public sphere. It materialized through an "interlocking of multiple networks and spaces" (Couldry and Dreher, 2007, p. 80).

Up to this point, the terms 'forum' and 'arena' have been used synonymously. In more precise understanding, though, 'forum' is more encompassing. According to Ferree et al. (2002, pp. 9-13), the forum is the core unit within the public sphere. It features three distinct parts: (1) the actual *'arena'* (sometimes also called 'front stage' in this research), (2) the *'backstage'*, and (3) the *'gallery'*. The following clarifies all three components of this stadium metaphor:

Arena/front-stage

The arena provides both individual and collective actors with a platform for voicing issues and opinions and introducing them into public discourse. In exchanging such speech acts, speakers may assume different roles, such as representing particular groups, providing specialist knowledge, or interpreting moral aspects (Peters, 2007, pp. 75-85). Some speakers may be professional spokespeople routinely participating in public discourse; others may only do so sporadically without special resources.

Depending on the nature of the forum, another role can also be found in the arena: that of mediators. Fora can be of various sizes and emerge on different levels – be it a quick chat among strangers at the supermarket check-out or nationwide public discourse sustained by mass media (see below). In some fora, the exchange of speech acts is not possible without technical means, i.e. media. In such cases, journalists are also present in the arena, observing speakers' contributions and making selections for what should be reported (see the concepts of *gatekeeping* and *agenda-setting*). At the same time, journalists may also provide context to and evaluation of speaker's positions (see, e.g., *framing*) and turn into speakers themselves. They do not only provide the stage, but may also climb it themselves (Pfetsch, 2008, p. 22). Which of the two roles is more dominant in journalistic routines depends on such factors as the type of media outlet or the national journalism culture (Hallin and Mancini, 2004; Hanitzsch et al., 2011).

Ferree et al. (2002) emphasize that arenas – especially those of comprehensive reach sustained by mass media – should not be imagined as stable, even, and well-lit turf. Instead, the terrain is full of obstacles whose "contours [...] can change suddenly [...] because of events that lay beyond the control of the players; and players can themselves sometimes change the contours through actions" (p. 62). These specific constellations of circumstances and contexts, which may bring about communicative advantages for some speakers and difficulties for others, are called 'discursive opportunity structures'. Speakers are advised to scrutinize the arena for such situations and adapt their contributions accordingly. "The activities and choices of the players, if they are to be successful, require that they be able to read this playing field and make their choices with an awareness of the opportunities and constraints that it provides" (p. 82). I return to this idea at a later point in this study, when discussing the short-term dynamics at the Cancún summit and their impact on actors' summit communication (see 3.3.2).

Backstage

As laid out earlier, processes of strategic communication entail both a concealed and a public part. Behind the scenes, communication *strategies* are crafted but commonly hidden from audiences' or other players' view. In the front-stage phase, these strategies are then carried out in the form of visible communication *activities*. The internal, strategic work takes place backstage, at an "organized production center" (p. 13), accommodating speakers' reading of the communicative terrain and the development of appropriate public discourse contributions. Here, speakers prepare before entering the arena and utilize dedicated resources put in place by actors for enhancing the effectiveness of their speech acts.

The lack of such infrastructure may constitute serious disadvantages in public discourse and the contest for audience approval (Feree et al., 2002, p. 13). Several accounts suggest a growing significance of the backstage for actors' public communication. Facilitated by such macro trends as the dissolution of overarching social institutions and milieus (*secularization*, *individualization*, *fragmentation*) and an increasing infiltration of media logic and scrutiny in many spheres of society (*mediatization*), actors are confronted with growing complexity when designing and implementing strategic communication programs (Blumler and Kavanagh, 1999; Kriesi, 2004). Navigating an actor through the intricacies of contemporary communication environments is the task of public relations (PR) professionals. Such holistic, strategic understanding of PR is put forward by Grunig and Hunt (1984), who see the field as "management of communication between an organization and its publics" (p. 6). Other definitions emphasize more selected aspects of PR, such as the running of press relations or the support of firms' marketing programs (e.g., Bentele, Liebert & Seeling, 1997; Szyska, 2009), or focus on specific concepts, like image or legitimacy (e.g., Merten, 1992; Raupp, 2011). However, such narrow definitions stay behind the more encompassing conception of PR as 'running the backstage' and reading – as well as responding to – the complexities of the arena's communicative terrain.

Despite its lack of an established corpus of professional knowledge, a widely shared ethos, or fixed training paths, the PR sector has undergone professionalization in recent years (Hahn, Mok, Roessler, Schmid, and Schwendeman, 2008, p. 337; Raupp, 2011, p. 102), visible for example in the launch of training institutions or an ongoing restructuring of the field into more refined functions, like graphic design creation, event management, or executive counsel. Already Hilgartner and Bosk (1988) identified "a whole sector of the economy that produces an ever-changing set of collective definitions of what we should be paying attention to and why" (p. 69) and that employs "operatives who specialize [...] in particular arena-based activities" (p. 68). Actors' backstage areas, to be found in corporate communications units, PR agencies etc., may feature a variety of structures and processes aimed at shaping their respective presentation in the arena.

Gallery

On the gallery, a third role in the public sphere can be found: the audience. Audiences follow the mediated or unmediated exchange between speakers and, through their virtual or physical attendance, turn such discourse into *public* discourse. The presence of an audience is a constitutive attribute of the public sphere – without observers, public-ness is nonexistent (Donges and Imhof, 2005, p. 155). Audiences remain mostly passive and abstract; they do not constitute an organized grouping, but heterogeneous, unspecialized aggregates (ibid.) without the capacity to act collectively (Feree et al., 2005, p. 13). In the case of public discourse via the mass media, individual audience members may take part through the occasional letter to the editor or 'vox pop' interview on the street (Gerhards and Neidhardt, 1991, p. 65), but other than that, audiences can only speak through particular collective actors, such as trade unions or pressure groups that assume speaking roles in the public sphere.

Despite its diffuse nature, the audience constitutes a key entity in public discourse. Its possible attention is taken into account when public speakers exchange speech acts (p. 44). Speakers' assumptions about audiences' likely expectations and reactions alter the character of discourse. The flashing of the 'on air' sign in a TV talk show may turn guests into enemies, although they are actually bound in friendship. The condition of public-ness instigates the addressing of a (virtual) audience and reinforces actors' interests and professional roles (Donges and Imhof, 2005, p. 154).

As suggested by the conception of the public sphere as intermediary system, (assumed) audience attention also plays a role in political decision-making. Demands that are expressed in front of a gallery full of political constituents are more likely to attain political outcomes than more quietly expressed claims. Public discourse is a source of orientation and persuasion for political decision-makers; especially what is written and broadcast for national audiences poses an influence (Ferree et al., 2002, pp. 14-15). 'Going public' has become a general term for actors' efforts of introducing their positions into public discourse and harvesting the influence of imagined audiences. Seen the other way round, this also means that "doing badly in mass media discourse creates vulnerability in pursuing policy interests" (p. 15).

The public sphere as network of different types of fora

The forum, consisting of arena, backstage, and gallery, is the basic unit of the public sphere. Contemporary public spheres comprise multitudinous fora of various types, which can be connected in different ways. Attempts in conceptualizing these links and describing the positions of various fora vis-à-vis each other have generated two dimensions of differentiation:

On a *horizontal dimension*, numerous less encompassing and more specialized fora cluster around the core of the national mass media forum. The latter constitutes a "master forum" (Ferree et al., 2002, p. 10), the backbone of nationwide public discourse, "*the* major site of political contest" (ibid., emphasis theirs). At the heart of this forum lies a small number of prominent news media, such as broadsheet newspapers, news magazines, or current affairs programs, which observe each other (Wessler, Peters, Brüggemann, Kleinen-von Königslow, and Sifft, 2008, p. 4). Some of their contents may circulate to or originate from other outlets of smaller reach that serve particular lifestyles, worldviews, or interests. These are sub-fora within the mass media forum, which may be partially influenced by what leading media report, but may also partially sustain separate public discourses (ibid.). The mass media forum is internally diverse; instead of a unified theater, it represents a multiplex venue comprising several auditoriums of various sizes – a few of them home to popular mainstream shows, but most of them housing more particular sideshows. And what is presented in one auditorium may well influence other shows under the same roof.

Besides the mass media forum and its respective sub-fora, public discourse also takes place in more specialized fora grouping around it (Ferree et al., 2002, p. 10). There, sets of speakers, mediators, and audiences are less encompassing and often defined by membership in an organization, profession, or other social group. A debate within an academic discipline or a political party's internal proceedings might be of such nature. The arenas of such specialized fora feature their own ensembles of speakers exchanging speech acts through dedicated media like trade publications or party papers. At times, public discourses within specialized fora may also enter the mass media through appropriate communicative linkages (Tobler, 2006, p. 110). The arguments of a debate in science might be introduced into the mass media forum (and stimulate public debate there) through science journalism. Similarly, a demand formulated in a social movement might enter public debate in the mass media via an NGO spokesperson, who is interviewed in a news program. And even the ordinary citizen, usually not holding a public speaking role, might appear in mass media discourse if his neighborhood initiative is featured in the city paper: "civic activities become part of the public sphere to the degree that they are represented in public communication, primarily in the mass media" (Peters et al., 2005, p. 140). Issues from all spheres of society are constantly fed into the mass media forum.

On a *vertical dimension*, Gerhards and Neidhardt (1991, pp. 49-56; see also Gerhards 1993; Neidhardt, 1994) distinguish different 'levels' of the public sphere. The different examples of fora that have been discussed up to this point constitute more elaborate constellations featuring a separation of (professional) roles and an abstract, non-physical setting in which speakers and audiences are connected mostly by technical means. However, public discourse in and around the mass media constitutes only one of three contexts, in which public spheres may emerge. On an ad-hoc level, public discussion can also occur on the street or at home – settings with spatial, temporal, and social boundaries (Donges and Imhof, 2001, p. 151). This is unmediated communication – simple systems of interaction (Gerhards and Neidhardt, 1991, p. 50) – in which speaking and listening roles are frequently switched and more private topics may take turns with those of public nature. Such episodes of communication are fragile and unstructured; they come into existence ad hoc, when individuals meet in the physical world, and may fall apart just as quickly.

More structured and topically focused is the type of public sphere that emerges at assemblies or protest events. Here, the roles of speakers and audiences are distributed more stably; selected individuals take the stage, whilst others form the audience. The latter may express direct approval or disproval with regard to what is happening on stage yet remains in a rather passive role. Mediators are usually not required at assemblies, due to attendants' physical presence. In terms of reach and distribution of fixed roles, assemblies are located between simple encounters and media discourse: "encounters on the street can be considered as the smallest unit of analysis while the mass media are the only forums which reach out to the broader public" (Brüggemann and Schulz-Forberg, 2009, p. 694).

Similar to the horizontal dimension, communicative linkages also exist vertically across these three levels. What is discussed in simple encounters may be carried into assemblies, which in turn might be noted in mediated public discourse. The other way round, issues debated in mediated discourse or at assemblies may be followed up by discussion in simple encounters. Viewed normatively, media discourse should be inspired by small-scale and medium-scale debate to stay relevant and authentic. Conversely, such discussions should be oriented towards media discourse to remain political in nature (Gerhards and Neidhardt, 1991, p. 56).

Communicative linkages as prerequisite for the integration of the public sphere

It is the horizontal and vertical communicative linkages between the different fora in the public sphere that make a good point for viewing it as one integrated domain. These perpetual flows of monitoring, processing, and distributing communicative content between different sites in society make up the public sphere (Habermas, 1992, p. 436). Only through these processes does the public sphere come into existence. Therefore, when contemporary public spheres are referred to as 'fragmented' (Brüggemann, 2008, p. 44; Brüggemann and Schulz-Forberg, 2009, p. 698; Latzer and Saurwein, 2006, p. 11), this certainly holds some truth if we take into view the 'infrastructure' formed by innumerable fora on different levels. Yet, the question of fragmentation relates more to the strength of the communicative linkages, the processes operating between individual components (Couldry and Dreher, 2007, p. 80): Are different sites of public discourse indeed open and attentive to each other? And to what extent do exchanges between various fora really justify the notion of one integrated network?

I have devoted so much space to this notion of the public sphere as a network of diverse fora integrated by communicative linkages because its serves as highly suitable foundation for understanding the transnationalization of public spheres (see section 2.3), which essentially is about nothing else than the emergence of communicative linkages crossing national borders. Analogous to the integration of sub-national or specialized fora in national public spheres, these connections are also imaginable for fora beyond the national level. But before we explore this idea more closely, I briefly touch upon the national context of public spheres.

2.1.2 The public sphere and the national context

As emphasized by its conception as intermediary system, the public sphere has commonly been modeled with an immediate view to the political system. Seen normatively, its incorporation, processing, and mediation of publicly relevant issues and opinions make it a crucial complement of political institutions in democratic societies (Gerhards and Neidhardt, 1991, pp. 39-41) and bring about the latent public approval, or legitimacy, that they require (Habermas, 2001b, p. 7; Sarcinelli, 2005, p. 81). Traditionally, such processes are organized nationally: Embedded in a "Westphalian frame" (Fraser, 2007, p. 9), citizenries enclosed by nation-states delegate their sovereignty and address their claims to national parliaments and governments, whose jurisdiction is limited by national territory. Besides elections, the primary sites for the definition of the political agenda are mass media fora, which, again, are mostly dominated by national outlets giving priority to issues of national relevance (Peters and Wessler, 2009, pp. 126-128). This segmentation is also facilitated by the fact that public discourse draws on what Peters (2007, pp. 106-116, 330-331) terms 'national public culture', a repertoire of shared assumptions, values, and symbols, including a shared language. While there may be diffusion and adaptation among national public cultures, some nationally distinct issues, strands of thought, or lines of conflict remain. As explored in the work of Karl Deutsch (1953, 1956), which is referred to in many parts of public sphere theory (Brüggemann, Hepp, Kleinen-von Königslow, and Wessler, 2009, p. 394; Koopmans and Statham, 2010a, p. 38; Wessler and Brüggemann, 2012, p. 63), this lets communication flows *within* national borders exhibit higher density than those *across* national borders. Nation-states hence constitute centers of gravity for public communication.

Nancy Fraser (2007) describes the national rooting of political institutions, mass media, and culture as the "implicit subtext" (p. 10) of traditional public sphere theory – and as fact that subsequent critique has failed to point out (p. 14). On various levels, Habermasian ideas are strongly linked with characteristics of the sovereign democratic state (pp. 9-10): Those who take part in the public sphere are seen as citizens, or members of a bounded *demos*, who possess the capacity of communicating in a common language and sharing something like a joint identity or experience. The outcomes of discussion are expressed as 'public opinion' and addressed to a sovereign state resting on an apparatus capable of enforcing political decisions. This discussion among citizens and the translation of demands into political decision-making is facilitated by national media, especially press and broadcasting.

2.2 Different perspectives on transnational public spheres

Considering the significance of the national context in classic public sphere theory, the question arises how the public sphere can be conceptualized to account for public discourse of transnational nature. Communicative spaces have, at least to some extent, become detached from national territories, thereby weakening – but surely not eradicating – the congruence of nation-state and public sphere (Brüggemann et al., 2009, p. 395). A process of recent decades, this development is linked to the growth of new technologies, such as satellite broadcasting or the Internet, and the emergence of intergovernmental or supranational political entities (ibid.) that de-territorialize political communication – even though the quality of these processes might be insufficient from a normative standpoint, as was laid out in the introduction.

Elaborations of this argument come from Fraser (2007) as well as Habermas (1998) himself, who sees the emergence of a 'postnational constellation', in which multinational corporations strip states of the capacity to effectively shape economic conditions (pp. 119-120) and governments delegate competencies to new supranational regimes lacking legitimacy (p. 108). Fraser adds that even though degrees of transnationalization could always be detected in public spheres (pp. 15-19), there has been an increase in the more recent past. Those engaging in public discourse do often not share equal citizenship, even if they reside in the same country. The omnipresence of "migrations, diasporas, dual and triple citizenship arrangements, indigenous community membership and patterns of multiple residence" (p. 16) has weakened the significance of national borders in deciding who may participate in such discourse and who may not.

Contrasting a dominantly national conception of the public sphere, several theoretical and empirical endeavors have attempted to take into account these transnational conditions of contemporary public spheres. However, scholarship has yet to agree on how to imagine a transnational public sphere – a fact that is visible in the variety of more or less elaborate theoretical approaches. These proposals can be lined up along a spectrum (see figure 2.1) between two poles representing distinct theoretical conceptions: While at one end of the spectrum, transnational public spheres are seen as distinct 'superstructures' detached from national public spheres, at the other end, they are looked at through national glasses and primarily perceived as transnationalized national public spheres.

2.2.1 Transnational public spheres as autonomous 'superstructures'

In some of the conceptions treating transnational public spheres as distinct domain or additional layer independently 'floating' above national public spheres, common features of national public spheres are translated to the transnational level, suggesting the existence of a separate space that is of similar composition than national public spheres. Castells (2008) draws on such perspective and sees "the same kind of common ideational ground that developed in the national public sphere" (p. 80) on a transnational level, located in the "political/institutional space that is not subject to any particular sovereign power" (ibid.). However, the exact nature of the structures and processes in this domain is "far from clear" (ibid.), as he admits. His notion of a transnational public sphere,[3] which appears to oscillate between normative and descriptive claims, is tied to his idea of a global 'network society' (1996), in which cross-border networks sustained or assisted by information technology permeate into all spheres of society: the net work as the core unit of human activity.

In an optimistic conception of a transnational public sphere at the global level, Volkmer (2003) draws on Castell's idea and suggests the existence of a colorful "spectrum of diversified transnational news flows" (p. 13) forming a global public sphere. As part of an "extra-societal" (ibid.) space, this transnational web of communication lies beyond national public spheres. Nonetheless, the

3 The terms 'transnational public sphere' and 'global public sphere' often appear to be used interchangeably in conceptions by Castells, Volkmer, and other theorists at this end of the theoretical spectrum. While particular geographic or social contexts for transnational public discourse are mentioned (such as ethnic diasporas or the Arab world), 'transnational public sphere', in these conceptions, is commonly regarded as globe-stretching.

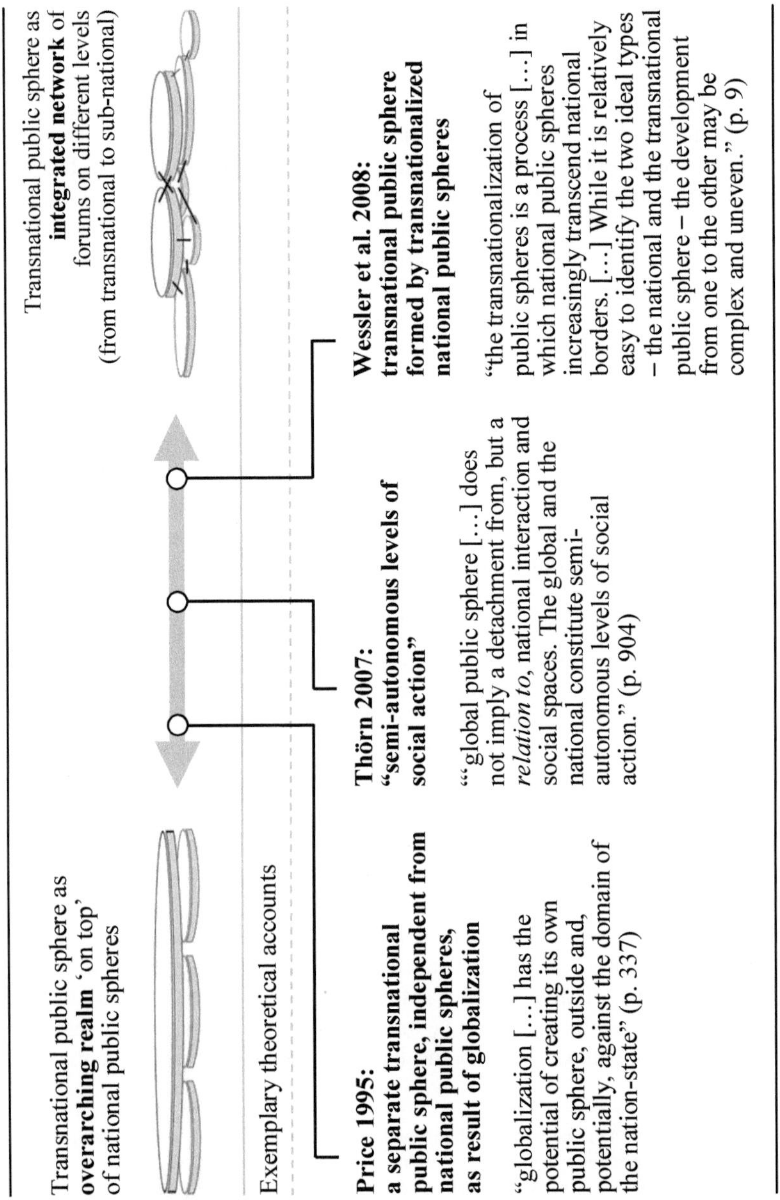

Figure 2.1: Contrasting theoretical perspectives on transnational public spheres

latter is "transformed by a new dialectical relationship between supra- and sub-national political contexts" (p. 15). Besides the network technologies of the Internet, transnational TV channels play an important role in Volkmer's conception. Particular emphasis lies on CNN International, which is described as one of the "worldwide dominant political 'hubs'" (p. 12), whose "role of a global authority has been widely underestimated" (Volkmer, 1999a, ¶ 5). Detached from – but with some influence on – national public spheres, the channel – but also other transnational TV channels – make up the global public sphere or at least become part of "transnational microspheres" (Volkmer, 2003, p. 13) around certain political issues. This is also the territory of global online discussion fora, news aggregation sites, or activist webzines. Indicative of a latent normative perspective, Volkmer presumes this global public sphere to provide communicative opportunities to marginalized political actors: National news agendas are opened to incorporate overlooked issues from around the world, and national censorship can be circumvented through "reciprocal communication" (p. 13). Also, particular transnational media products may increase actors' global visibility. For instance, CNN International's *World Report* granted stations from less visible locations the opportunity of having their reports distributed across the globe without any interference (Volkmer, 1999b):[4] "Through this unique format and CNNI's worldwide distribution, new political perspectives and new 'players' entered the global sphere, […] who are able to participate in this new transnational political sphere and present authentic political perspectives for a global audience which they would not reach otherwise" (Volkmer, 2003, p. 12).

Strong references to particular media products as agents of transnational public spheres can also be found in Lull's (2007) euphoric conception. He sees a "global commons" (p. 157) formed by electronic and digital media. They carry "public discourse that transcends national borders and creates diverse and democratic participation in the global public sphere" (ibid.). Blogs, for instance, scrutinize the statements of political and economic actors regardless of their national origin, forming "system-correcting mechanisms on a global scale" (ibid.). Chat and VoIP programs train people around the world in free speech, even if restricted in particular countries. And new transnational channels like Al-Jazeera open up spaces for discussion stretching across the world. As part of a development trajectory envisioned by Lull – "the seven stages of the open spaces of global communication" (p. 151) –, these developments may result in new levels of global consciousness and "global wisdom" (p. 162), since "the very size of the global audience empowers the persons who form it" (p. 161). Worldwide mobilizations against apartheid in South Africa, the United States' invasion of Iraq, or

4 This refers to *World Report* in its old format; at the time of writing, a generic news program was broadcast under that name.

the detainment of Burmese politician Aung San Suu Kyi are cited as corresponding examples. The strong normative ambitions of this conception are clearly visible in its overall promotion of "meaningful dialogue and nurturance of the global public sphere" (p. 169).

Individual technologies or media outlets as transnational public spheres?

It becomes apparent that authors like Castells, Volkmer, and Lull – I chose these three as mere representatives of this perspective – regard transnational public spheres as detached from national public spheres – in line with Price's (1995) statement that "globalization [...] has the potential of creating its own public sphere, outside and, potentially, against the domain of the nation-state" (p. 337). While these authors point at interconnections and relations of influence between both levels, the lens through which they analyze communication beyond national borders is largely void of national contextualization. Put differently, in this view, theorizing of transnational communication occurs without much theoretical attention to national communication. Both levels are seen as autonomous: National public spheres primarily sustained by national mass media are juxtaposed with transnational public spheres of more or less global scope, which are formed by transnational infrastructure like transnational TV stations or Internet platforms. It is a fundamentally *dichotomous* outlook.

Conceptually, this particular way of modeling transnational public spheres appears to lack some analytical precision; it often remains unclear *how exactly* such conceptions can guide the understanding of public discourse across national borders and could be operationalized for empirical studies. Cottle and Rai (2008) detect only "limited empirical engagement" (p. 163) in this stream of research and argue that it "advances suggestive but often speculative claims" (p. 164). Especially proposals overburdening single technologies or outlets with euphoric hopes of creating an integrated commons or global civil society should be greeted with skepticism. Also empirically, the attribution of a key role in the creation of transnational public spheres to selected media outlets or Internet platforms remains problematic. Sparks (2005), for example, acknowledges the rise of internationally distributed or truly transnational media outlets, such as *The Financial Times*, CNN International, and prominent Internet sites: "It is tempting to claim that these developments represent at least the foundations of a global public sphere" (p. 38). However, measured against criteria of reaching global mass audiences and allowing for the legitimization of global governance through broad public discourse, a different picture emerges, as these outlets mostly reach "a predominantly male, well-educated and well-off group of people" (p. 42) in the West, which should not be mistaken for a comprehensive global audience.[5] Moreover, the contents provided by such outlets are less transnational in nature than commonly assumed. Transnational TV channels, for example, depend on national regulators, funders, and infrastructure, which is often reflected in their programming; also, they often feature sub-networks customized for different markets.

And also the Internet – although frequently hyped – cannot be seen as the backbone of a global public sphere, as it is mostly absent outside the West and some "pockets of wealth in the developing world" (p.44). The sheer lack of hosts – or electricity, for that matter – makes any discussion of "a global public sphere [...] meaningless in such situations" (ibid.). Through an analysis of online fora tied to social movements and transnational initiatives, Cammaerts and van Audenhove (2005) investigate in more detail whether empirical evidence justifies the notion of an emerging transnational public sphere formed by the Internet. They, too, come to a sobering conclusion: "While the issues being addressed may be transnational, participants are often located in the Western hemisphere, discussion often happens between likeminded activists [...] restrained by language and cultural barriers" (p. 194).

It could be argued, however, that the explored sites represent specialized fora and that nowadays, there exist some true 'transnational mass media' in the online world. Above all, Facebook, which was still small at the time of Cammaerts and van Audenhove's study but meanwhile approaches the frontier of one billion members (Facebook, 2012), qualifies as candidate for investigation. The platform sees occasional episodes of focused communication that, through mechanisms not always clear, result in a wide-ranging viral 'buzz' cutting through the otherwise rather fragmented interactions on the platform – see the example of 'Kony 2012' in early 2012, when a US charity's online video publicizing the crimes of Ugandan warlord Joseph Kony went 'viral' and was looked at 70 million times in four days (Visible Measures, 2012).

Maybe such young phenomena of high-attention episodes come closer to what can be seen as transnational public spheres in the online domain. This, however, would still be different from the stable, all-encompassing communication infrastructure of global reach that enthusiastic theorists see in the Internet. In terms of scope and inclusion, it is a far cry from constituting a separate transna-

5 Lull (2007) himself acknowledges CNN International's small elite audience: "CNN International appeals mainly to global middle-class residents, tourists, and business travelers. CNN, BBC International, and other international broadcasters transmit by satellite and cable, whose subscription costs lie beyond the economic grasp of the vast majority of the world's population" (p. 163). However, Lull argues that the stories broadcast on these channels commonly 'trickle down' into other outlets as well as interpersonal networks.

tional public sphere – let alone from satisfying the particular qualities of public discourse that normative theorists would look for. Hence, individual outlets like Facebook, CNN International and others might play an important role in connecting and influencing public spheres around the world, but regarding them as exclusive constituents of transnational public spheres seems conceptually simplistic and empirically unjustified. Essentially, "the public sphere is, and always will be, a much larger phenomenon than an Internet discussion forum" (Agre, 2002, p. 311).

2.2.2 Transnational public spheres as outcome of transnationalization

If we approach the middle of the theoretical spectrum that I introduced above, we find perspectives on transnational public spheres that, on one hand, acknowledge the existence of national public spheres but, on the other hand, still treat both domains as essentially distinct. In these 'middle positions', both levels are linked in one way or the other. For instance, Thörn (2007) argues that "'global public sphere' does not imply a detachment from, but a relation to, national interaction and social spaces" (p. 904). Both levels are "semi-autonomous" (ibid.); they are interwoven with each other, but still remain separate structures; hence, transnational public spheres are not just the "sum of a number of national [...] public spheres" (p. 903). Braman (1996) sees "multiple, overlapping and interpenetrated public spheres" (p. 34) that are located on different levels. While the national level remains important – and the state influential in shaping public discourse (p. 35) –, multinational corporations and NGOs are active in transnational public spheres interpenetrating national ones (p. 27).

At the very end of the spectrum, we find conceptions that are not too far away from Braman's view. Here, authors propose to abort the dichotomous perspective differentiating between transnational and public spheres and advocate a conception of transnational public spheres that "builds on the existing national public arenas, viewing them as the institutional hardware" (Lingenberg, 2009, p. 47). Especially researchers empirically investigating particular transnational discourses, such as those in particular world regions (e.g. Europe) or surrounding particular issues (e.g. climate change) theorize transnational public spheres with strong attention to their manifestation in national public spheres. They essentially look at the transnational *through the lens of the national* and thereby challenge the theoretically abstract and empirically problematic perspectives of 'free-floating' transnational public spheres of global scope. I return to the matter of transnational public spheres in Europe and surrounding climate change at a later point (see 2.3.1); however, it is not only scholars working on these topics who have propagated a theoretical perspective on transnational public spheres incorporating national public spheres.

National public spheres as the substance of transnational public spheres

Olesen (2005), for example, regards national public spheres as the substance of transnational public spheres, although the former are certainly transformed by what is happening on the transnational level. These transformations originate from "transnational publics" (p. 424), which are discursive spaces around certain issues or events that stretch across national borders and are often facilitated by social movements – an understanding that is conceptually not far away from what was defined above as 'fora'. Speakers and mediators active in the forum formed by national mass media might pick up on such transnational discourses and thereby introduce them into national public spheres. The transnational public sphere hence denotes the variety of transnational publics *as made visible in national public spheres*. Only through this translation to the national level does the transnational public sphere come into existence: "Without this physical visibility in national public spheres, transnational publics could still be transnational, but they would be public only in a limited sense" (p. 433). Interactions with the transnational level can also be found elsewhere on the national level besides the mass media; Olesen points to topically focused fora at the assembly level, like those formed by NGOs, which are often influenced by transnational discourses. The transnational public sphere hence stands for new ways of "combining the local, the national and the transnational" (p. 420), but it is rooted in the mediated and face-to-face interactions in the national public sphere.

A similar argument is formulated by Couldry and Dreher (2007) based on the case of community media tied to transnational ethnic groups in the 'global city' of Sydney. Investigating the communicative linkages between these outlets and the mainstream forum of Australian mass media, they propose to see such specialized niche fora not "simply as counter-public spheres operating in parallel to a unitary mainstream public sphere, nor as local public 'sphericules' floating unattached to any shared space of dialogue" (p. 96). As "each is potentially contributing over the longer term to a larger space of dialogue with the mainstream" (ibid.), they instead advocate a view in which particular transnational fora (represented by Sydney community media) may connect with national or sub-national fora (represented the Australian mass media).

And indeed, such perspective makes the normative notion of counter-public spheres superfluous, since it does not regard the public sphere as one unified space anyway, "but as the interlocking of multiple networks and spaces" (p. 80) on various levels: sub-national, national, and transnational. Hence, the focus shifts to the relations between these various fora and – from a normative viewpoint – to their stability and equality (Asen and Brouwer, 2001). According to the arena model, there is no reason to assume that transnational fora are excluded from such webs of communicative linkages. Just as any sub-national forum might be integrated in these networks, this can also apply to transnational ones. While the nation-state remains the prime structuring unit for the public sphere and national mass media represent centers of gravity for public discourse, national fora may open up for transnational ones. This is a "gradual and multidimensional process" (Brüggemann and Schulz-Forberg, 2009, p. 695) that moves public spheres to higher – or lower – levels of transnationalization.

It becomes clear that at this end of the spectrum, the theoretical perspective on the transnational public sphere differs from the dichotomous outlook of the euphoric theorists cited above. What the latter regard as an *(semi-)autonomous structure* parallel to national public spheres is seen as particular *quality* or *state* of national public spheres by others. Both archetypal perspectives take into view the same phenomena (public discourse stretching across national borders, partially assisted by transnational outlets), but when it comes to the incorporation of the national level, the theoretical contextualization differs. While one camp sees only little relation between national and transnational public discourse and hence focuses on the latter, the other camp does not regard this as "either/or discussion" (Hepp and Wessler, 2009, p. 175, own translation) or "yes or no question" (Peters et al., 2005, p. 141). These scholars view the process not as the *emergence of transnational public spheres*, but as the *transnationalization of national public spheres* (Wessler et al., 2008, p. 8).

2.3 Transnationalization of national public spheres

Two avenues for the transnationalization of national public spheres figure prominently in the literature: (1) *gradual, long-term* restructuring of public spheres and (2) *episodic, short-term* orientation towards a particular event (Brüggemann et al., 2009, p. 408; Peters and Wessler, 2006, pp. 139-140). The "gradualist" (Peters and Wessler, 2006, p. 139, own translation) or "structural" (Wessler, 2009, own translation) model takes into view how the public sphere's ensemble of fora and actor constellations slowly change towards a transnational quality. This approach is rooted in a *discursive* communication perspective, which views communication as speakers' exchange of speech acts. Transnationalization, in such perspective, refers to measurable characteristics of public discourse. In contrast, "episodic" (Peters and Wessler, 2006, p. 140, own translation) or "situational" (Wessler, 2009, own translation) approaches focus on the ad-hoc impact of exceptional events on national media and their audiences around the globe. Here, we are in the domain of *ritual* communication, of scrutinizing public communication for celebratory, community-reaffirming elements (see 2.3.2).

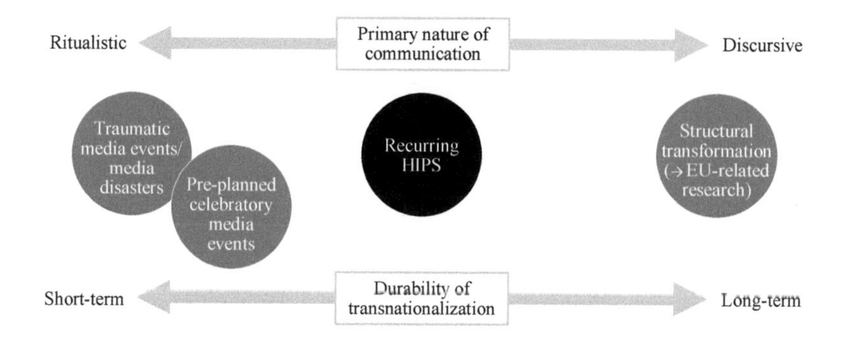

Figure 2.2: Theoretical approaches to the transnationalization of public spheres

While the two perspectives emphasize different aspects of communication (exchange of information vs. facilitation of ceremony), they are applicable to the same communication processes and contents. It is my argument here that high-level international political summits (HIPS) unfold transnationalizing potential through both mechanisms (see figure 2.2). They may contribute to a structural transnationalization of national public spheres, especially if they recur regularly (like the UN climate summits under investigation here), but they may also come close to performing the function of internationally noticed 'media events' (see 2.3.2), facilitated by the attachment of symbolic resources by various stakeholders, including those actors from civil society that enjoy access to these events. I go into the notion of HIPS and their defining and optional features in more detail below; for the time being, the gradualist/ structural and episodic/situational avenues of transnationalization are explored separately.

2.3.1 *Long-term restructuring of discursive communication*

Before exploring research on the transnationalization of national public discourses on the issue of climate change, which is in the focus of this study, it is important to understand how this long-term restructuring of communication can be conceptualized.

2.3.1.1 Operationalizing transnationalization

As mentioned earlier, gradual restructuring of national public spheres towards higher degrees of transnationalization has extensively been studied in the context of the EU. Researchers working on this topic have investigated to what extent the growing political integration between EU member states is associated with communicative integration, i.e. Europeanization (as one particular kind of transnationalization) of national public spheres. Similar processes are imaginable for other territorial frames of reference. Schäfer et al. (2011) list world regions that can be conceptualized as foundations for regional transnational public spheres. Besides the case of the EU, where political integration suggests the emergence of stronger communicative links within the confederation than with the outside, investigations into the transnationalization of public spheres make sense for regions bound by a common language, for example the Arab world (Ayish, 2008; Lynch, 2003) or Latin America (Cañizález and Lugo-Ocando, 2008; Sinclair, 2005).

However, long-term transnationalization processes may not only take place in particular world regions but can also occur through interaction with deterritorial phenomena, such as social movements stretching across national borders (Beyeler and Kriesi, 2005; Stammers and Eschle, 2005), diasporic communities (Fazal and Tsagarousianou, 2002; Hepp, Bozdag, and Suna, 2011), or long-standing issues relevant to more than just one country[6] (Beck, 1997, Seitz and Hite, 2012). Hepp (2006, 2009) lists four forms of deterritoriality – he sees them as 'cultural thickenings' – supplementing and overlapping with territorial settings for condensed communicative interaction: ethnic (diasporas), commercial (pop cultural communities), political (social movements), and religious (religious communities): "[T]he borders of the cultural thickenings people belong to do not necessarily correspond with the territorial borders, while at the same time territories still have a high relevance as a reference point" (2009, ¶ 13). Hepp argues that assisted by developments in information and communication technology, such deterritorial spaces for thickened communication have blossomed in recent times. Be it the cross-border anti-globalization movement or Star Wars fan community – national (or other territorial) borders have lost some of their structuring power in the face of such phenomena. While not all of these spaces are home to political communication and public discourse (Hepp discusses them

6 Such issues are often subject to campaigns by transnationally operating NGOs embedded in social movements. Hence, deterritorialization facilitated by social movements and by global issues should not be seen as fundamentally different phenomena.

mainly from the perspective of overlapping cultures and identities), some can certainly be regarded as transnational fora carrying the potential of structurally transforming national public spheres. National fora, such as the master forum of the mass media, might open up for transnational fora through such processes and generate communicative linkages between both levels.

Such structural transformation of public spheres towards higher levels of transnationalization cannot be detected by looking at one place only. Instead, more than one indicator has to be taken into account, several dimensions considered (Brüggemann et al., 2009). Wessler and Brüggemann (2012, pp. 64-68) put forward a four-dimensional heuristic for the detection of transnationalization in public spheres. Accordingly, the phenomenon might be visible in the deterritorialized orientations of (1) technical and social *infrastructures* of public spheres, i.e. media outlets and professionals, (2) *speakers* and their back-stage strategizing, (3) *audiences* and their reception patterns, as well as in (4) *public discourse* itself. While these dimensions are closely intertwined, they are often studied separately – just as in this research, where the focus is on speakers' transnational strategies (dimension 2) with implications on the cooperation with media professionals (dimension 1), the manifestation of public discourse (dimension 4), and perceptions held by audiences (dimension 3).

Within the body of literature on a possible transnationalization of public spheres in the EU, some studies have explored aspects of production or reception (dimensions 1-3) and, for this, drawn on observations, surveys, or interviews, as in the research by Brüggemann (2008) on the political PR of the European Commission, Raeymaeckers, Cosijn, and Deprez (2007) on the routines of national journalists based in the 'European capital' of Brussels, or Lingenberg (2009) on citizens' perceptions of the failed referenda on an EU constitution in 2005. The transnational quality of public discourse itself (dimension 4) is measured in content analyses. Regarding Europeanization, this has been done within some larger research projects, such as Koopmans and Statham (2010b) or Wessler et al. (2008). The latter employ four dimensions for measuring transnationalization of public discourse, although one should not expect these dimensions to feature explicit 'tipping points' dividing national from transnationalized degrees (pp. 21-22). Instead, change on these dimensions is a matter of gradual development and "structural transformation" (p. 9). The dimension of (1) *monitoring governance* refers to national media's coverage of institutions and policies of governance above the national level, (2) *discourse convergence* to possibly emerging similarities among national public discourses, e.g. in terms of problem definitions or discourse alliances, (3) *discourse integration* to the attention paid to political developments in other countries or the exchange between speakers in different countries, and (4) *collective identification*, finally, to, e.g., the expression of belonging to a transnational community in public discourse (pp. 10-12). Some of these dimensions can be captured by looking at one country only. But as soon as we are interested in transnationalization phenomena of particular scope (e.g., European or global), we need to put national public spheres side by side and assess whether findings across countries paint the picture of a transnational public sphere based on transnationalized national public spheres.

2.3.1.2 Transnationalization around the issue of climate change

For the issue of climate change, which is in the focus here, such multi-country comparative approaches have rarely been taken on a wider scale. Schäfer et al. (2011) analyze leading print media in 23 countries in all world regions (with the notable exception of South America) and look for similarities in the level and trend of national climate change discourses (see the 'discourse convergence' dimension described above). While they generally find low levels of media attention to the issue in all countries, they also detect a common increase of coverage around the years 2006 and 2007. What can also be found across countries are more short-term fluctuations in the level of coverage; such peaks are influenced by high-visibility events like the 2009 Copenhagen climate summit. Statistically, media attention to the issue is clearly more similar between certain countries than among others. Especially within Europe, but also across the Atlantic (Europe, US, and Canada), national climate change discourses develop in a similar way. Such similarities cannot be found for other regions. Based on these findings, the authors see "no evidence for the existence or emergence of a global public sphere" (p. 133) around the issue of climate change.

Eide, Kunelius, and Kumpu (2010) also look at climate change coverage around the world through the lens of a potential global public sphere. Based on a content analysis of elite and popular newspapers in 19 countries on all continents, they investigate the level of reporting devoted to the Copenhagen summit as well as the nature of voices cited. They also examine the frames conveyed in coverage, although this is done separately for each country on the basis of open questions and not synthesized in a truly comparative manner. In general, the authors see increased media attention to COP-15 across countries – "a global high point" (Eide and Kunelius, 2010, p. 21). This peak was partially due to journalism's deliberate attention-building in order to "champion the success of the negotiations" (p. 41). But despite this "attempt at a global public sphere" (p. 42), coverage was strongly adapted to national priorities and perspectives, hindering the emergence of common public discourse across national borders. In absence of precise comparative indicators for the degree and quality of transnationalization, this is what we are left with as overall conclusion from this study – besides a general corroboration of the findings by Schäfer et al. (2011) that climate summits can stimulate media attention around the world.

The orientation of national discourses towards the issue of climate change has also been studied for individual countries. As a matter of fact, case studies of various aspects of specific national discourses, especially those in Western societies, dominate this field of research (Schäfer, Ivanova, and Schmidt, 2012, p. 122). While some of these studies contain comparative perspectives (see, e.g., Boykoff, 2007 [US/UK]; Brossard, Shanahan, and McComas, 2004 [US/France]; Dirikx and Gelders, 2009, 2010 [Netherlands/France]; Shanahan, 2009 [newly industrialized as well as non-industrialized countries]; Boykoff, 2010 [20 countries on all continents]), they often lack methodical soundness to be truly comparative. Also, many of them are purely descriptive and require explanatory context (Schäfer et al., 2012, p. 123).

Owing to the diversity of these approaches, it is difficult to deduce wider conclusions for the transnationalization of national climate change discourses from this literature. However, particular patterns in climate change discourses around the world can be detected (see also Dirikx and Gelders, 2008; Wessler, 2012). For instance, differences exist with regard to the prominence of climate change skeptics and the representation of (the rather low degrees of) scientific uncertainty attached to climate change findings. Boykoff and Boykoff (2004) point to the US, where the journalistic norm of balanced reporting leads to a consideration of skeptic voices that bears no proportion to their marginalized status in science, although this might have improved in the recent past (Nisbet, 2011). Still, Zehr (2000) detects scientific uncertainty as "a highly salient theme" (p. 98) throughout US newspapers' climate change coverage, attributable to the portrayal of controversy and the procedural nature of scientific research. Carvalho (2007) shows for UK newspapers that the representation of skepticism and uncertainty is not shared homogenously but related to the paper's ideological orientation. And even more contrastive, Weingart, Engels, and Pansegrau (2000) find "remarkable consensus" (p. 281) on the existence of anthropogenic climate change in German print media: "scientists politicized the issue, politicians reduced the scientific complexities and uncertainties to CO_2 emissions reduction targets, and the media ignored the uncertainties and transformed them into a sequence of events leading to catastrophe" (p. 280). Inspired by such findings, Dirikx and Gelders (2008) speculate whether the media treatment of climate change may differ more systematically between the US and the EU.

Differences between national climate change discourses also become visible in the attribution of responsibility. Here, a conscious delineation by non-industrialized or newly industrialized countries vis-à-vis industrialized countries becomes visible. The Indian press, for example, clearly sees the latter as causers of climate change and hence views mitigation regimes with mistrust, namely as efforts by the West to limit India's economic growth (Billet, 2010). This is a "nationalistic reaction in the press that reinforces the non-compliance of India and its public with any binding emissions targets" (p. 15). Similar frames are found in the Chinese press, which include references to the country's low per capita emissions, despite its first rank in absolute emissions (Midttun, Coulter, Gadzepko, Wang, and Staurem, 2012). This "legitimates its claim to a different climate-responsibility than the rich West" (p. 24). Outside BRICS societies, in non-industrialized countries, levels of media coverage on climate change are generally low (Anderson, 2009, p. 169). Besides a scarcity of resources, this is due to journalists' low levels of subject expertise and overload with current developments and debates; this "confusion has made it difficult for journalists and editors to report with any confidence about how climate change will impact their communities" (Kakonge, 2011, p. 1).

Empirical knowledge on patterns of transnationalization around the issue of climate change is hence limited. While research has provided some first hints at groups of countries sharing similar manifestations of public discourse (Europe vs. the rest of the world; BRICS countries vs. industrialized countries), detailed findings regarding transnationalization are limited. What we do know, however, is that UN climate summits can serve as universal triggers for media coverage and hence become subject to simultaneous observation around the world.

2.3.2 Short-term episodes of ritualistic communication

The attraction of worldwide media attention to particular events has also been explored in a different stream of literature focusing on so-called 'media events'. For a discussion of the concept, we briefly have to put on hold the perspective of discursive communication employed thus far. Media events, instead, concern the ritual aspects of communication and bring to the fore what McQuail (2010) has subsumed under the 'ritual model' of communication: "Ritual or expressive communication depends on shared understandings and emotions. It is celebratory, consummatory (an end in itself) and decorative rather than utilitarian in aim and it often requires some element of 'performance' for communication to be realized" (p. 71). An analysis of ritual communication scrutinizes speech acts, which are exchanged in public fora, for particular symbols and performances "invoking and sustaining public solidarities based on ideas and feelings (collective sentiments) about how society should or ought to be" (Cottle, 2006, p. 416). Ritual functions of mediated communication can be detected for routine and exceptional circumstances. Carey (2009) provides the everyday example of reading a newspaper, which, in a ritual view, is seen "less as sending or gaining information and more as attending a mass, a situation in which nothing new is learned but in which a particular view of the world is portrayed and confirmed" (p. 16). The concept of media events, however, is concerned with extraordinary situations of ritual communication.

The starting point of the theoretical discussion of media events is Dayan and Katz' seminal study *Media Events: The Live Broadcasting of History* (1992). In this book, integrating communication studies with anthropological approaches, they describe media events as the "high holidays of mass communication" (p. 1) and as specific genre of television, primarily performing ritual functions. Archetypes of media events can be found in the Olympic Games, the first moon landing, or royal weddings. In ideal form, they can be regarded as a force of social integration, prompting a comprehensively unifying form of attention and appreciation on the side of broadcasters and audiences. In particular, Dayan and Katz define media events by eight criteria:

- They are *planned in advance*. They do not happen unexpectedly but are made possible by an often elaborate process of organization.
- They are *planned outside the media*. While they become media events only through the help of the media, preparations and core performances are mainly done by other actors.
- They *monopolize media attention*. They are broadcast by several channels, with almost no alternative programming taking place.
- They are *broadcast live*. Viewers know that the occurrences displayed on television are taking place at that very moment.
- They are *presented in a formal, respectful, and sacred manner*. Dayan and Katz speak of "reverence and ceremony" (p. 7) and "reconciliation" (p. 8) surrounding the broadcast.
- They *interrupt daily routines*. People stop doing what they normally do in order to be able to follow the event on television.
- They *monopolize audience attention*. The broadcast seizes very large audiences, which have almost no chance of escaping it.
- They *prompt festive viewing*. People apprehend the historic nature of the moment, paying tribute to it as such.

While all of these attributes have to be present in order for a media event to constitute the distinct TV genre envisaged by Dayan and Katz (Hepp and Couldry, 2010, p. 2), they can follow three distinct scripts, or patterns of dramaturgy: contest, conquest, and coronation. *Contests* are high-stakes "rule-governed battles of champions" (Dayan and Katz, 1992, p. 26) eventually yielding a winner. *Conquests* see "heroes pushing back a frontier, overcoming ostensible laws of nature and society" (ibid.) and thereby changing the world. *Coronations*, finally, are all about ceremony, involving respect for authority and a cultivation of tradition. Drawing on Max Weber's three forms of authority – rationality (contests), charisma (conquests), and tradition (coronations) –, the authors conceived these scripts as frames of analysis through which all media events could be studied (Hepp and Couldry, 2010, p. 2-3).

Different conceptions of media events

Dayan and Katz's classic notion of media events appears to originate from Cold War times of political constancy and limited-channel television; hence, the question emerges how much validity the concept possesses in contemporary times. In fact, Dayan (2010) himself offered a reworked version, in which he elaborates on some of the more recent challenges to the original conception. Firstly, in their semantic value, media events do not anymore draw on the overarching theme of conflict and peace; they are no longer "gestures that seemed to lessen the possibility of war" (p. 26). Secondly, the exclusive TV genre of media events has been tattered. Its elements can now be found across the media landscape, as media outlets and audiences have a harder time agreeing on what qualifies as media event. This "multiplication of 'almost' media events leads to the emergence of a 'gray' zone" (p. 29) positioned between media events and routine news coverage. Thirdly, while classic media events involved a "machinery of suspension" (p. 28), including the sole focusing on the broadcast, deferral of parallel activities, and loyal acceptance of the event's main definition, today's media events face strong competition from other influences: "a constant invitation to disengage from the surrounding community" (p. 27) and facilitator of "individualized reception" (p. 28). This contradicts the loyal communal celebration of such events envisioned in the original concept.

Various scholars have critically received and extended the original concept of media events. Cottle (2006) surveyed various types of what he calls 'mediated rituals' – defined as "exceptional and performative media phenomena that serve to sustain and/or mobilize collective sentiments and solidarities on the basis of symbolization and a subjunctive orientation to what should or ought to be" (p. 415). For him, media events in the tradition of Dayan and Katz constitute only one of these types, "celebratory media events" (p. 418). Another type is called 'media disasters' and based on a contribution by Elihu Katz himself, together with Tamar Liebes (2007). They do not reserve the label 'media event' for preplanned, ceremonial, and integrative events but extend it to include surprising, shocking, and disruptive happenings. "For the fact is that media events of the ceremonial kind seem to be receding in importance, maybe even in frequency, while the live broadcasting of disruptive events such as Disaster, Terror and War are taking center stage" (p. 158). These three types of occurrences – together with Protest (including Revolution), which is also mentioned but not elaborated on (p. 161) – can be regarded as the traumatic counterparts to the three ceremonial scripts by Dayan and Katz. They are covered on television through 'disaster marathons' (Liebes, 1998), which are different from journalism's normal 'bulletin mode' in that they dwell on the event for hours or even days, expressing shock and repeating the same footage, evaluations, and speculations. In this sense, just as ceremonial media events are co-produced by organizers and broadcasters, the media also play an essential role in disruptive, negative media events. Whenever such events are preplanned, e.g. by anti-establishment actors or terrorists, the 'construction support' provided by media is clearly anticipated by such actors (Katz and Liebes, 2007, p. 164).

Cottle's set of mediatized rituals has been criticized for its overambitious collapsing of various phenomena into one grand category at the expense of analytical value (Couldry and Rothenbuhler, 2007; Hepp and Couldry, 2010). And indeed, not all conceptions that circulate under the label 'media event' or are presented as conceptually related to media events actually carry an event focus. Hepp and Vogelgesang (2003), for example, argue that in popular culture and consumer marketing, various 'popular media events' can be found. While Katz and Liebes' traumatic media events still share many of the original Dayan and Katz criteria, Hepp and Vogelgesang are more far off. They speak of an increasing 'eventization' of media, detectable, for instance, in heavily advertised 'blockbusters' (often promoted as 'TV events' by the channels themselves) or cross-promoted sports matches or pop concerts. While such events may also interrupt daily routines, they do so in a much milder way. They are not necessarily broadcast live, emphasize pleasure, only affect some media segments (and, hence, only excite and integrate certain audiences), and are organized by the media themselves.

Media events and the performance of symbolic resources

In the context of this research, conceptions of media events that also include occurrences without clear spatial and temporal boundaries are less relevant. In the focus here are those physical events at which media carry out a "*sine qua non* function" (Kunelius and Nossek, 2008, p. 255, emphasis theirs): only through extensive honoring by the media (be it in a reverent or appalled manner), which is anticipated and prepared for by organizers and others, does the event become a media event, an object of ritual communication. This aspect of media 'performance' (Hepp and Couldry, 2010, p. 12) is a necessary condition; occurrences that are not sufficiently performed by and in the media are no media events. What makes a performed event different from one that is simply reported on is the attachment of 'symbolic resources' (Wessler and Brüggemann, 2012, p. 13): "symbols, narratives, and cultural codes [...] which citizens can experience contemporaneously with everyone and interpersonally with those around them" (Alexander and Jacobs, 1998, pp. 27-28). If these symbolic resources are simultaneously performed by media in multiple countries (if the performance of the event is hence not only limited to one country) and met with heightened degrees of attention and corresponding ritualistic appreciation (even if only in mild form), we may attribute some transnationalizing effects to such events.

However, we have to be careful not to overvalue the integrative potential of transnational media events; there is no linear, monocausal link to the national level. Media events can almost never be seen as stimulus for the emergence of a "common 'we'" (Hepp and Couldry, 2010, p. 12), but rather as triggers "for the construction and reconstruction [...] of many varied national, ethnic, religious, subcultural and other voicings of that 'we'" (ibid.). They constitute reference points, or backdrops, for domesticated discourses that integrate the event with national frames and "master narratives" (Kunelius and Nossek, 2008, p. 268). This is visible in media reporting, but also on the side of audiences. While Kyriakidou (2008), investigating audience reactions to the Southeast Asian Tsunami and Hurricane Katrina, detects "rare moments of expression of global solidarity" (p. 288), she generally concludes that the response to these events should be seen as fluid and fragmented, structured by "dispersed multiple connections across various socio-geographical levels" (ibid.). Media and audiences in different parts of the world cannot be expected to homogenously share interpretations of media events (see also Wessler and Brüggemann, 2012, p. 115).

Symbolic resources as target of strategic action

The interpretation of media events can also be subject to strategic action. Dayan (2010) regards them as "exploited resource[s] within a political economy of attention" (p. 28), as "strategic venues" (ibid.). Organizers/perpetrators may attempt to compose a dominant "thematic core" (Hepp and Couldry, 2010, p. 11), or central message, for the event, articulated through purposefully chosen symbolic resources. This, however, maybe met with various efforts of contestation and appropriation around the world. Challenging the attribution of uniform ritualistic effects, Cottle (2006) regards media events as "productive spaces for social reflexivity and critique" (p. 411) that also allow marginalized or counter-establishment actors to get heard with their event interpretations. This essentially leads to Dayan's (2010) pointed question: "Can anyone own a public event?" (p. 30). He argues that media events may

> "fall prey to entities that are neither organizers nor their publics. They may be subverted (denounced), diverted (derailed), or perverted (hijacked). They may be used as Trojan horses or placed under the threat of a sword of Damocles. These multiple tensions and the calculated moves of various public actors interested in the exploitation of the event's charisma ask the question of 'legitimate ownership' and undue appropriation" (ibid).

While this implies a somewhat binary perspective on the contestation around media events ('good' organizers on one side, 'bad' challengers on the other), it does point us to their quality as communicative resources and domains for appropriation. This is taken up in the empirical chapters below, where I investigate actors' attempts to craft the symbolic resources that media can draw on in their ritualistic performance. I do not argue that the climate summits under investigation here should be seen as classic media events in the tradition of Dayan and Katz. However, they can certainly be seen as providers of symbolic resources (some more than others) facilitating transnationalizing effects. This symbolic loading of HIPS is explored in more detail in the next chapter (see 3.2.2).

2.4 The summit as transnational forum and provider of symbolic resources

It is my argument here that the transnationalizing capacity of HIPS rests in two mechanisms:

For the period of the summit, public discourse, as represented by national mass media, is temporarily altered towards the inclusion of contributions by summit actors. The summit can be seen as distinct *short-term forum at the transnational level*, which on the one hand accommodates public discourse for itself (i.e. summit-internal exchanges between actors as part of negotiations or via 'summit media' like *Earth Negotiations Bulletin* or *ECO* – see 6.2.5) and on the other hand interconnects with other forums, such as those sustained by national mass media in different corners of the world. According to conceptions of the public sphere as flexible network made up by various public fora on different levels and the communicative linkages within and between them (see 2.1.1), there is no reason why we should not regard summits as such fora interlocking with other fora – be it at the transnational, national, or sub-national level. For a better understanding of the variety of interconnections between summit discourse and national discourses, it makes sense to remember Wessler et al.'s (2008) transnationalization dimensions (see 2.3.1.1).

At the same time, a summit can draw strong attention worldwide through its perceived symbolic significance. Symbolic loading (visible in the attachment of, e.g., visuals and slogans, links to culture and history, or references to 'grand narratives') can be conducted as deliberate strategy by hosts, media, or other actors or might emerge from preceding public discourse. As *provider of symbolic resources*, a summit might approach what was introduced above as 'media event' (see 2.3.2), although the classic criteria of this genre are probably never achieved by any contemporary summit. In the context of this research, it suffices to focus on the role of media and their 'performance' of the symbolic resources made available by stakeholders. Media's ritualistic honoring may facilitate the summit's worldwide reception and thereby promote transnationalizing effects.

The distinction between both mechanisms draws on the conceptual differentiation between two models of communication (communication as exchange of information vs. communication as carrier of rituality). However, both mechanisms may be visible in the same summit coverage and work with mutual reinforcement: Statements in public discourse may contribute to a summit's symbolic loading, which in turn may reinforce public discourse around the event.[7] In this integrative perspective, which simultaneously regards a summit as locus of public discourse of global relevance and as provider of symbolic resources, the full transnationalizing potential of these events rests in an entanglement of both mechanisms.

7 The effects of short-term episodes of media attention to climate summits on a long-term transnationalization of public discourse have become subject to empirical investigation (Wessler, 2009; Wessler and Adolphsen, 2011).

3. Summits as strategic communication resources

The second theoretical chapter of this study deals with the role of summits as resources for strategic communication. I first examine prior research on strategic political communication on the international stage (3.1). Preceded by some general insights regarding summits, I then clarify my conception of high-level international political summit (3.2). In a third section, both aspects are combined in an exploration of research on strategic summit communication (3.3.1). Based on all literature reviewed in the theoretical chapters, I finally develop a conceptual model that explains actors' choice of summit communication strategy (3.3.2).

3.1 Public diplomacy: strategic communication on the international stage

Strategic political communication efforts that involve a crossing of national borders are often labeled as 'public diplomacy'. While the term was already used in Cold War times to refer to communicative tools employed by the two blocs in their confrontation (Gregory, 2008) –visible, e.g., in US-run Radio Free Europe or the episode of 'Ping Pong Diplomacy' between China and the US –, scholarly interest in the concept increased after 9/11, when the need for coordinated processes of international communication appeared urgent. In the US, the attacks drew attention to the country's bad image in some parts of the world and triggered debate on communication mistakes on the international stage as well as possible adjustments (see, e.g., Van Ham, 2007; Vlahos, 2009). Beyond the case of the US, the concept has been discussed in the context of an ever-more complex international system featuring new actors, advanced communication technologies, and a growing importance of public opinion in foreign relations (see, e.g., Gilboa, 2001; Melissen, 2007, Price, 2009).

Traditionally, public diplomacy has been viewed as the domain of states. As such, it is often associated with the notion of 'soft power'. In contrast to coercion and reward as established means of (military and economic) power in the international arena, soft power rests in a country's capacity as target of sympathy and respect, as role model in the world (Nye, 2008, p. 94). This relates to such intangibles as character, values, or conduct – or, to be exact, to how they are perceived in other countries. Soft power is hence a matter of subjective social construction and, as such, strongly interwoven with processes of (strategic) communication. In clear contrast to realist thinking in international relations theory (see, e.g., Waltz, 1979), adherents of public diplomacy have argued that soft power has become "more important than territory, access, and raw materials, traditionally acquired through military and economic measures" (Gilboa, 2008, p. 56) in the international system. In other words, "countries that are likely to be more attractive in postmodern international relations are those that help to frame issues, whose culture and ideas are closer to prevailing international norms, and whose credibility abroad is reinforced by their values and policies" (Melissen, 2007, p. 4). Kunczik (2003) even sees the possibility of 'image fights' between countries, which may also bear consequences on hard power (p. 135), visible, e.g., in a drop in foreign direct investments. In this thinking, soft power, as a country's exploitation of "attraction and seduction" (Nye, 2008, p. 95), may precede or facilitate some forms of hard power.

3.1.1 Definitional issues

According to a pragmatic definition, public diplomacy refers to "an international actor's attempt to advance the ends of policy by engaging with foreign publics" (Cowan and Cull, 2008, p. 6) or, similarly, "the process by which direct relations with people in a country are pursued to advance the interests [...] of those being represented" (Sharp, 2007, p. 106). In the traditional understanding of the soft power perspective, the actors of public diplomacy are governments of sovereign states carrying out a two-step process: by influencing publics in other countries, they eventually influence their respective governments. Correspondingly, Snow (2009, p. 6) argues that the field of diplomacy has seen a gradual transformation: Conventional government-to-government ('G2G') interactions, run by envoys and embassies hidden from public sight, were widened to include a 'detour' via foreign publics and became common *public* diplomacy ('G2P'). With the rise of new communication technologies and a growing importance of public opinion in foreign affairs, however, non-state actors, e.g. from business or civil society, also started to conduct public diplomacy ('P2P'). This final phase has also been termed '*new* public diplomacy' by some scholars (see, e.g., Melissen, 2007; USC Center on Public Diplomacy, 2009).

The notion of 'new' or 'P2P' public diplomacy should be treated with a grain of skepticism, however. On the one hand, it seems plausible to adjust conceptions of public diplomacy to contemporary circumstances and abandon the definitional limitation to state actors. In a world in which "governance occurs above, below, and around the state" (Gregory, 2008, p. 283), political communication is no longer the exclusive domain of states. Instead, "[l]arge and small non-state actors, and supranational and sub-national players develop public diplomacy policies of their own" (Melissen, 2007, p. 12). Accordingly, the conception of 'actors' in the definitions above should be seen as holistic. Be it a multi-country appeal by UNICEF against the use of child soldiers, a cross-border anti-famine effort by Médecines Sans Frontières, or an EU-wide PR campaign by Nestlé advocating free trade – provided that such communication programs (1) *relate to the shaping of political conditions* and (2) *involve a communicative crossing of national borders*, they should be counted as public diplomacy.

On the other hand, the definitional expansion of the notion of public diplomacy should not be taken too far. Some authors suggest a rather loose understanding, which could also stand for international communication as a whole. For example, Snow (2009), referring to the rise of social media stretching across national borders, directs our view to "the various publics and diplomacies that are engaging, collaborating, combating, and just bumping into each other" (p. 8). Mueller (2009) even regards the "web of human connections" (p. 102) between countries as infrastructure of a public diplomacy carried out "one handshake at a time" (ibid.). Cull (2008) also highlights the value of interpersonal ties, regretting that "asylum seekers and recent migrants are not generally seen as a public diplomacy resource but merely a welfare problem to be managed" (Cull, 2008, p. 50). While such everyday interactions among individuals may indeed be relevant for the image and policies of international actors (in fact, this is the idea behind state-run foreign exchange services, such as the Fulbright Program in the US, or cultural relations institutions, like the German *Goethe-Institut*), they lack the element of (3) *strategic orientation*. Yet, this must be regarded as integral feature of public diplomacy; it is meant to persuade and influence (Dearth, 2002, p. 4; Fisher and Bröckerhoff, 2008, p. 6) and hence constitutes goal-oriented communication. The arbitrary cross-border interactions of individuals – except maybe of those occupying professional public diplomacy roles – cannot sensibly be included in such a strategic communication perspective (Paul, 2011, p. 36).

In order to account for the strategic character of public diplomacy and for the fact that this usually involves political actors' (4) *use of mass media* for the distribution of their messages, Entman (2008) puts forward the term 'mediated public diplomacy'. This refers to "shorter term and more targeted efforts using mass communication (including the internet) to increase support of a country's specific foreign policies among audiences beyond that country's borders" (p. 88). With the exception of its limitation to state actors, this definition comes closest to the view adopted here. If we assemble all relevant definitional characteristics discussed up to this point, we arrive at a definition of public diplomacy as state and non-state actors' mass communication efforts (including those on the Internet) ultimately directed at the shaping of political conditions and at least partially targeting audiences in other countries than the country of origin.

In accordance with Entman's definition, mass communication here is not regarded as the exclusive task of established 'offline' mass media; it may also take place via online services. Also, the focus on actors' ambitions to concretely shape political conditions does not rule out long-term objectives, such as image-building, which, in the long run, are also driven by political ambitions. Finally, this definition applies to public diplomacy of both international and transnational nature (see 1.3), as it also covers strategic communication efforts that are concurrently directed at audiences in several countries, including maybe the country of origin.

While some conceptions of public diplomacy carry more particular normative connotations – visible mostly in an emphasis on dialogue and peace-making (see, e.g. Melissen, 2007; Riordan, 2004) –, the definition developed above remains on a broad descriptive-analytical level. Of similar nature is the definition of PR presented earlier (see 2.1.1). And indeed, there have been proposals to treat public diplomacy and international political PR as essentially identical, since both terms denote an actor's strategic cross-border communication and hence refer to the same organizational function, even though differences might exist with regard to concrete objectives and instruments (L'Etang, 2009; Wilcox, Ault, and Agee, 1992). With governments, IGOs, NGOs, and corporations engaging in similar strategic communication on the international stage, calling these processes 'public diplomacy' for some actors and 'international political PR' for others would be conceptually incoherent. I recognize this conceptual overlap (Signitzer and Coombs, 1992) by treating both notions as synonymous.[8]

3.1.2 Previous research: governments' public diplomacy

Besides definitional work, the domain of traditional, i.e. 'G2P', public diplomacy has been strongly neglected in theoretically grounded research, as several scholars have noted. Gilboa (2008) detects a predominance of US-focused historical case studies with little theoretical value in the public diplomacy literature. Similarly, Snow and Taylor (2009) complain that "empirical data and reasoned analy-

8 On a side note, Snow (2009) adds that many of the scholars and pundits advocating a clear difference between public diplomacy and PR simply do not want to see public diplomacy suffer from the allegedly bad reputation of PR (pp. 9-10). Such politically motivated differentiations, however, cannot be justified conceptually.

sis from academic schools of thought are often overlooked in favor of perfunctory opinion editorials and discourse from a narrowcast of retired generals and diplomats" (p. ix). However, some first steps have been taken to compensate this shortage of "theoretical infrastructure" (Entman, 2008, p. 87) in the literature on public diplomacy, such as attempts at typifying different forms or components of public diplomacy.

Typologies of public diplomacy

For Cull (2008), the practices of public diplomacy can be divided into five elements: (1) *listening*, the gathering of information on the attitudes of foreign publics; (2) *advocacy*, the implementation of PR work in the target country; (3) *cultural diplomacy*, the spread of a country's cultural resources beyond national borders, for example through institutions like *Instituto Cervantes* or the British Council; (4) *exchange diplomacy*, the sending abroad of domestic citizens while receiving foreign citizens, for example in the context of study-abroad programs; and (5) *international news broadcasting*, the dissemination of television and radio programs tailored for specific foreign audiences, for example by Voice of America or *Deutsche Welle*. The problem with this taxonomy is the overlap between its categories and the apparent confusion between functions and tools of public diplomacy. For example, the sponsoring of an exhibition by domestic artists touring foreign countries could be counted as cultural diplomacy but, at the same time, is inherently linked with advocacy, if not listening.

Gilboa (2001) proposes a different conceptual structure for coming to grips with the wide range of public diplomacy activities. He regards public diplomacy as only one of three distinct constellations of the use of the media in international relations. Besides public diplomacy, these are *media diplomacy*, the use of the media by foreign affairs officials to communicate directly with state actors in other countries (without the targeting of broader audiences) and *media broker diplomacy*, i.e. the active involvement of journalists in bridging hostilities and conducting negotiations. *Public diplomacy*, now, refers to "direct communication with foreign peoples, with the aim of affecting their thinking and ultimately, that of their governments" (p. 4). This may happen in three ways (see figure 3.1):

In the *basic variant*, mass media and other means of communication are used for reaching the minds of people in other countries. This activity is about creating "a favourable image for a country's policies, actions, and political and economic system" (p. 5). The assumption is that "if public opinion in the target society is persuaded to accept that image, it will exert pressure on its government to alter existing, hostile, attitudes and policy" (ibid.). For Gilboa, a context of hostility between the targeting and the targeted country seems indeed to be a defining feature of this model, as he assigns its use mainly to Cold War times. As a simple activity of advocacy, however, it is still used today.

The *domestic public relations variant* draws a more complex picture of the process of influence, as it involves the services of third-party actors, namely of public relations firms in the targeted countries. They are hired by the initiators of public diplomacy efforts for two reasons: Firstly, PR agencies in the targeted country may bring in comprehensive knowledge essential to an effective dissemination of messages in that society. Secondly, they may help conceal the sources and financing of communication efforts, as it might not always be in the interest of governments to be known as conductors of public diplomacy. Due to its incorporation of professional expertise and PR techniques, Gilboa also calls this variant "strategic public diplomacy" (p. 7). However, this labeling is not shared here, as public diplomacy is assumed to be always strategic (see 3.1.1).

When it comes to the *non-state transnational variant*, a basic assumption of older conceptions of public diplomacy is dropped. Here, not only national governments attempt to influence foreign publics. All kinds of other political actors are also active in the international arena, each drawing attention to particular issues or advocating specific interests. These communication activities also have to be counted as public diplomacy efforts. In contrast to conventional public diplomacy, however, they are not always targeted at one particular country but aim at generating broader international support in several societies at the same time. This commonly works through the staging or exploitation of events drawing worldwide attention – an aspect that I will return to below.

Beyond typologies, theoretically nourished studies of public diplomacy are rare. First efforts in empirically investigating public diplomacy based on theories from communication studies are portrayed in the following.

Theoretically rooted investigations of public diplomacy

As an extension of his 'cascading network activation model', Entman (2008) makes a proposal for how the framing of US foreign policies in other countries' media may be shaped. This is to explain the level of success of US mediated public diplomacy, measured as degree of parity between US government-endorsed frames and foreign media frames. He outlines a complex chain of factors that all bear influence on the framing of US policy in foreign media – among others, the opinions expressed by foreign elites or the public's ability and motivation to consult alternative news sources (pp. 96-97). In general, Entman theorizes, attempts at the influencing of media coverage as part of public diplomacy can unfold most comprehensively where media operate freely in a context of pluralism. In these societies, media framing may be mixed but generally in line with the degree of cultural congruence between that country and the US. ('Cultural congruence' refers to shared values, experiences, and outlooks.) However, where media are centrally controlled, the leeway for public diplomacy is smaller, as it is restricted by the ruling elite's (positive or negatives) attitudes towards the US (p. 97). Efforts by the US administration (or any other government, for that matter) to exercise control over how its foreign policies are covered abroad must thus be seen as highly contingent processes.

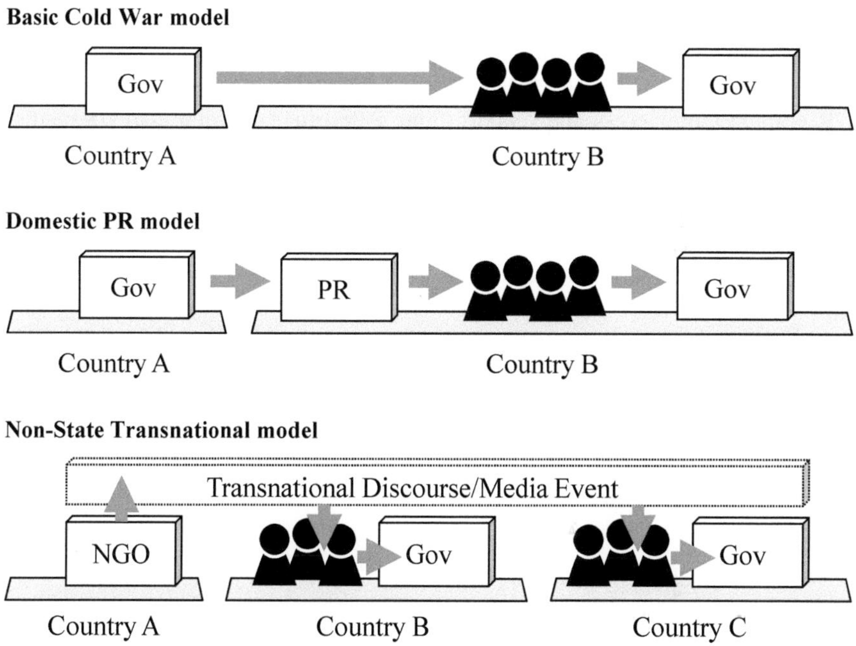

Figure 3.1: Three variants of public diplomacy (modeled after Gilboa, 2001)

Focusing on strategic international agenda building and strategic international frame building, Sheafer and Gabay (2009) pick up on Entman's notion of cultural congruence as prime factor facilitating the spread of desired (i.e. public diplomacy-induced) frames into foreign coverage. For the two events of the 2005 Israeli withdrawal from parts of the Palestinian Territories and the 2006 Palestinian general elections, they investigate the degrees of attention and support UK and US media (newspapers and TV networks) grant to Israeli and Palestinian positions and what frames they feature. The underlying assumption is that Israel and the US share some elements of cultural congruence, as do the Palestinians and the UK, and that differences in coverage between the UK and the US must reflect this: “the greater the cultural resonance between two countries, the more a government will successfully ‘push’ its frames into the second country’s media” (p. 451). The authors discover that during both events, the respective event initiators were able to promote their agenda and trigger the desired media fallout in the culturally closer media system (p. 455): “the more the initiated event correlates with news media values (e.g., dramatic, emotional, and negative events), the more the initiator succeeds in promoting its agenda” (p. 463). Attention to and support for Israeli officials was highest during the Israeli event – likewise for Palestine officials during the Palestinian event (p. 456). Generally, media in the UK granted more access and support to the Palestinian position than US media. Attention to and support of the Israeli position was essentially similar between UK and US media, with the exception of one phase of investigation (post-Israeli disengagement), when US media showed a bias towards Israel. In terms of framing, Israel succeeded more in getting its frames into the US media than into the UK media – and the other way round for the Palestinians (despite some more nuanced limitations).

Both contributions stress the significance of cultural and political congruency for successful public diplomacy and also point to the considerable influence by officials and journalists in target countries on how another country’s policies are covered. While both insights stem from explorations of traditional public diplomacy originating in country A and targeting country B (see Gilboa’s ‘Basic Cold War model’), they are certainly also applicable to transnational public diplomacy.

3.1.3 Previous research: NGOs’ public diplomacy

An exploration of what has been written about the public diplomacy of non-state actors – particularly of environmental NGOs (ENGOs), which are in focus here – brings about a mixed picture of the state of research. In general, the number of studies on the roles of NGOs in international environmental politics is considerable (see, e.g., Doyle, 2009; Gough and Shackley, 2001; Jamison, 1996; Johnson and McCarthy, 2005; Keck and Sikkink, 1998b; Newell, 2000; Princen and Finger, 1994; Raustiala, 2001; Yearley, 2003). Within this literature, it is widely recognized that ENGOs constitute established players in processes of global environmental governance, although this does not mean that they are legally integrated in such procedures. Sikkink (2002) sees some general tension between the widespread “acceptance of the importance of NGOs in international life, and the meager formal recognition of and provisions for their influence in international institutions” (p. 304), although it has to be noted that the UNFCCC, whose summits are under investigation here, is known for being relatively accessible to NGOs (see 3.2.3). What NGOs lack in formal access and control they have to compensate by means of communication, turning them into archetypal soft power players (p. 303). The instruments of soft power that are discussed in the literature are diverse, ranging from lobbying politicians to provision of technical expertise to implementation of pressure campaigns. Public diplomacy, though, is about the latter means, about *publicly visible* acts of strategic communication across national borders.

If the focus of literature review is narrowed down accordingly, the range of applicable studies gets smaller. While many studies deal with ENGO campaigning in a domestic context (see, e.g., Rucht, 1999; Voss, 2007; or, for a detailed overview, Schmidt, 2012), research explicitly dealing with their *cross-border* communication efforts (i.e. with public diplomacy rather than political PR) is limited. The international or transnational work of ENGOs is mostly discussed in *sociological social movement literature*, which deals with the circumstances under which social movements emerge as well as their societal functions and effects (Khagram, Riker, and Sikkink, 2002) and, within international relations, in *global governance literature*, which treats NGOs as counterparts or supplements of states in international policy-making (Take, 2007). In both strands of research, the aspect of strategic public communication beyond the national level is commonly sidelined or even ignored altogether; concepts and approaches compatible to communication studies are rare.

In a notable exception, Keck and Sikkink (1998a) study the transnational advocacy work of NGOs in the fields of environment, human rights, and violence against women. I briefly outline four general approaches that these authors detect in NGOs' cross-border communication efforts and use this typology for integrating additional findings from other pertaining literature. By and large, NGOs' selections from this repertoire of approaches are shaped by their overall mission and ideology, according to some studies. Doyle (2009), for instance, detects a division between institutionalized ENGOs and loosely organized environmental grassroots groups, which also has effects on preferred communication strategies. Similarly, Diani and Donati (1999) categorize four general types of NGOs: those that aim at the creation of conventional pressure and those that prefer more disruptive approaches; both may rely on professional or participatory/voluntary resources for implementation. Thus, how NGOs make use of the approaches laid out in the following is also a question of organizational purpose and culture (see also Voss, 2007, pp. 41-44).

As part of (1) *information politics*, NGOs treat information as strategic resources and connect their dissemination with specific strategic objectives (Keck and Sikkink, 1998a). Such information are often associated with accounts of individuals suffering from harsh conditions caused by powerful others (a farmer having to cope with severe draught, a revolutionary being tortured by a dictatorial regime etc.). The value of NGO information commonly rests in this character as individual testimonies, “stories told by people whose lives have been affected” (p. 19). While this allows for the element of drama that is commonly appreciated by media and enhances chances of media distribution, NGOs have to be careful not to jeopardize their status of providers of trustworthy and detailed information (ibid.). The facts have to be right, since only accuracy turns NGOs into respected players – but facts that no one pays attention to are also worthless. Correspondingly, Voss (2007) identifies some tension between political urgency and PR suitability as factors shaping ENGOs’ information politics but generally concludes that ENGOs are proficient in making the balance of facts vis-à-vis emotions match the targeted outlet or audience (pp. 272-273). Newell (2000) highlights that climate change is a complex issue to communicate, as it provides little room for drama à la ‘good vs. bad’ or ‘right vs. wrong’ (p. 125). And as NGOs commonly strive to be on the ‘good’ side, to operate on the “moral high ground” (de Jong, 2005, p. 120), this complexity might pose problems for communication purposes. Generally, establishing “connections between distant global processes and the effect of everyday actions” (Newell, 2000, p. 125) is a particular challenge in NGOs’ information politics.

Similar to the differentiation between discursive and ritualistic communication introduced above (see section 2.3), ENGOs supplement information politics with (2) *symbolic politics*, which are to add force to their factual messages. This can happen through spectacular images or memorable slogans suggesting particular symbolic readings or interpretations of ‘what this all means’ (Keck and Sikkink, 1998a, pp. 22-23). The crafting of symbolic resources by ENGOs has a long tradition. Especially ‘image events’, ‘photo ops’, or ‘stunts’ are a frequently used tool in the repertoire of symbolic politics (Delicath and DeLuca, 2003; DeLuca, 2001; Greenberg, 1985). Such actions may feature the physical confrontation of an opponent (like the typical blockade of a power plant) or a humorous performance ridiculing a political leader; their essential attributes are in any case visual appeal and message clarity. Especially Greenpeace has made headlines with such ‘direct actions’ in the past, using them to ‘bear witness’ to the misdemeanors of environmental culprits and thereby create images with strong symbolic meaning (Warkentin, 2001, p. 66). Such activities are also appreciated by audiences for their embodiment of unselfish risk-taking; activists become “heroes in an age of very few heroes” (Princen, 1994, p. 35). But also outside stunts,

NGOs craft “powerful visual imagery to ‘sell’ climate change” (Gough and Shackley, 2001, pp. 338-339). These images often depict environmental disasters and climate extremes – a tactic whose long-term educative value is questionable (ibid.; Doyle, 2007).

Information and symbolic politics can be used for purposes of (3) *leverage politics*, which consist in the strategic selection of targets of advocacy with the intent of increasing its efficacy (Keck and Sikkink, 1998a, p. 23). The underlying assumption is that certain actors (e.g., governments or corporations) are more prestigious or powerful than others and that a policy change carried out by such actors is more likely to be imitated by others than if carried out by less prominent actors. Also, NGOs may assume target actors to be more responsive to persuasion attempts by third parties and thus engage in coalition-building (ibid.). The general idea is to increase ‘leverage’, to set up routes of strategic communication processes with an explicit view to increasing chances of initiating policy change. A similar exploitation of hierarchies and mutual observation among actors can be found in (4) *accountability politics*, where NGOs use information and symbolic politics for drawing attention to actors’ failure to stick to public commitments. This is also known as ‘public shaming’, the highlighting of the “distance between discourse and practice” (Keck and Sikkink, 1998a, p. 24), and may be particularly effective with actors overly concerned about their public image.

ENGOs’ transnational public diplomacy as ‘two-level gaming’

While all four approaches can also be carried out in a domestic context, Keck and Sikkink conceive of them as commonly involving the communicative crossing of national borders. The global scope of some issues has shifted the centers of political decision-making, which are the common targets of advocacy, to a transnational level, while some governments also shield off domestic NGO pressure and thereby promote the use of advocacy ‘detours’ via the transnational level (p. 12). The latter constellation is called ‘boomerang pattern’ (pp. 12-14); “domestic NGOs bypass their state and directly search out international allies to bring pressure on their states from outside” (p. 12, see figure 3.2). A similar pathway of NGOs’ public diplomacy is found in the ‘spiral model’, which Risse and Sikkink (1999) developed for describing how repressive states slowly respond to opposition or NGO pressure exercised internationally (via liberal states) or transnationally (via IGOs). On the issue of human rights, which they investigate, this may be visible in gradual concessions and the ultimate adherence to international human rights norms by the target state (pp. 17-33). ‘Boomerang pattern’ and ‘spiral model’ hence suggest particular routes for the strategic communication approaches described above. Especially with regard to increasing leverage and enforcing accountability, the exploitation of international and transnational channels might increase the effectiveness of NGOs' strategic communication.

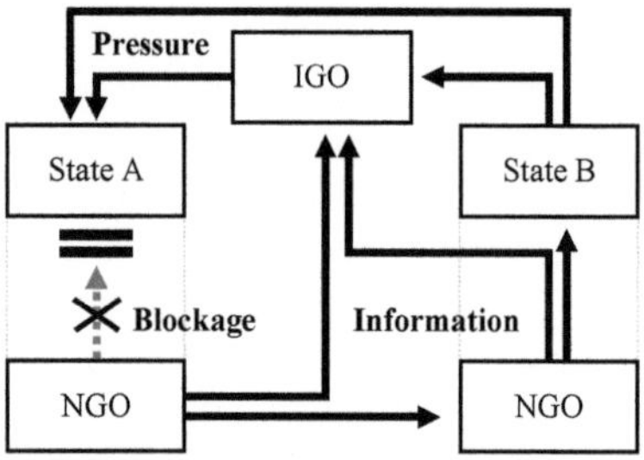

Figure 3.2: The 'boomerang pattern' in NGO public diplomacy (modeled after Keck and Sikkink, 1998a, p. 13)

While Newell (2000) acknowledges the value of 'going transnational' in ENGOs' advocacy, he stresses the significance of national strategic communication. Especially with regard to agenda-setting, i.e. efforts "to politicize an issue that was not previously considered political in an overt sense" (p. 129), national politicians are still primary target actors, as it is them who decide upon the issues negotiated in international institutions and hence determine the overall political dynamic in these fora (p. 162). However, at the stage of actual negotiations, international and transnational communication processes certainly play a role, visible in ENGOs' coordinated lobbying of politicians in multiple countries (pp. 138-139) or campaigning at summits (pp. 145-146). Finally, strategic ENGO communication that aims at making actors stick to their publicly made commitments may put to use a combination of domestic strategies drawing on established relations with national political actors (p. 151) and transnational shaming of governments (p. 149). ENGOs' strategic communication hence combines domestic, international, and transnational processes in various constellations; ENGO public diplomacy should not be seen as being fully detached from national advocacy, but rather as part of flexible 'two-level gaming'[9]. Even though international institutions have become shared targets in strategic communication efforts by NGOs worldwide (Rucht, 2001) and "coordinated international campaigns [...] against international actors, other states, or international institutions"

9 See Putnam (1988) for the classic political science concept of 'two-level games'.

(della Porta and Tarrow, 2005, p. 3) have evolved as part of this, the national circumstances of these phenomena as well as connected processes of national advocacy cannot be ignored (pp. 10-11).

3.2 Conceptualizing high-level international political summits (HIPS)

Linked to my discussion of HIPS as transnational fora and providers of symbolic resources (see section 2.4), I now clarify what I mean by 'high-level international political summits'. Providing an answer requires some elaboration on the concept of the summit and its features. While some authors trace the roots of modern-day summitry back to tribal congregations of ancient times (Dunn, 2004, p. 137) and royals' consultations in the Middle Ages (Melissen, 2003, p. 6) and see it as a phenomenon "as old as human history" (Weilemann, 2000, p. 16), the practice of using the term 'summit' for referring to political meetings is not even a century old. Confronted with the rising tensions of the Cold War, British premier Winston Churchill demanded the upholding of dialogue between countries, proclaiming that "it is not easy to see how things could be worsened by a parley at the summit" (Dunn, 2004, p. 138; Melissen, 2003, p. 2; Reynolds, 2007, p. 1). The notion of the political summit was hence born out of a context of confrontation, in which "rivals, if not antagonists" (Dunn, 2004, p. 138) met to "discuss issues of high politics of global consequence" (ibid.). In such classic understanding, the summit is seen as the domain of experienced statesmen, as an intimate setting for the shaping of political order across entire continents.

Defining HIPS

The (1) *participation of high-level politicians* is the most frequently noted definitional feature of summits. In strict understanding, these can only be serving heads of state and/or government or the political principals of IGOs (Melissen, 2003, p. 4). Other politicians holding cabinet rank may also be involved in summits but must enjoy a comprehensive executive mandate in order for the summit to constitute "diplomacy at the highest possible level" (Dunn, 2004, p. 148). In a traditional view, the agreements concluded by officials at a summit are not to "be second guessed by any other individual" (ibid.). Additionally, summits are usually seen as (2) *physical face-to-face meetings*; high-level contacts conducted via phone or video link are not counted (Dunn, 2004, p. 149; Weilemann, 2000, p. 17). Additional characteristics vary among authors, but commonly stipulated is (3) *some form of topical focus or substantive purpose*. While summits often carry strong symbolic value and entail appropriate formality (such as opening ceremonies, military rituals, or presentation of flags), purely ceremonial events, like weddings, funerals, or many official 'state visits', are not included (Melissen, 2003, p. 5). A summit is, at least partially, a "business meeting" (Reynolds, 2007, p. 5) with a focus on policy-making; it must include at least one pre-planned session (a negotiation round, an exchange of views etc.) aimed at generating policy-relevant results. This is not to say that summits always contribute to substantive policy-making – many authors actually argue for the opposite (see, e.g. Weilemann, 2000, p. 18) –, but the event must at least be set up in a manner that such results *could* emerge.

Beyond these criteria, 'the summit' is a rather loose concept. Summits come in various forms, which can be seen in the "wide gulf between the intimate *tête-à-têtes* of Allied leaders during the Second World War and the global summits at the beginning of the twenty first century" (Melissen, 2003, p. 4). Some authors put forward typologies of summits – differentiating between the, e.g., 'personal summit', 'plenary summit', and 'progressive summit' (Reynolds, 2007, p. 425) or the 'serial summit', 'ad-hoc summit', and 'exchange of views' (Berridge, 1995, p. 83) –, but such lists are neither exhaustive nor mutually exclusive. Others point to the significant differences between bilateral and multilateral summits and regard the field of multilateral summits alone as diverse and lacking an "ideal type" (Fomerand, 1996, p. 362). For the multilateral environmental negotiations under the auspices of the UN, Seyfang (2003) distinguishes between occasional "mega-conferences" (p. 224; see also Clark, Friedman, and Hochstetler, 1998) covering numerous issues (like the 1992 'Earth Summit' in Rio de Janeiro) and recurring summits that are more topically focused, like the COPs of the UNFCCC under investigation here – although the scope of topics negotiated at COPs (the 'climate summits') has also expanded strongly (Hunter, 2010). To keep matters simple, I combine the three definitional features just described in one pragmatic definition of HIPS as

> physical meetings of limited duration bringing together politicians of at least cabinet rank from two or more countries for the purpose of generating policy-relevant output.

Three optional features of HIPS (two of which are explored more fully below) are particularly interesting in the context of this research: As discussed above, a summit may move toward the status of a media event. This relates to the (4) *attachment of symbolic value* by stakeholders, i.e. journalists 'performing' the summit and promoting its symbolic resources. In this study, these symbolic resources are assumed to facilitate the transnationalizing potential of HIPS; their prominence and construction are hence noteworthy aspects, which I return to in the empirical chapters. For the time being, it should suffice to say that COP-16 did certainly not enjoy as much "fanfare" (Hunter, 2010, p. 4) as the preceding Copenhagen summit but still was the carrier of substantial symbolic resources.

<table>
<tr><th>Defining features of HIPS</th><th>Optional features of HIPS</th></tr>
<tr><td>Participation of politicians holding at least cabinet rank from two or more countries</td><td>May become object of symbolic loading/carrier of symbolic resources</td></tr>
<tr><td>Physical, face-to-face meeting of limited duration</td><td>May involve actors from civil society</td></tr>
<tr><td>Planned with a view to the generation of policy-relevant output</td><td>May come in form of a summit series with regular recurrence</td></tr>
<tr><td colspan="2">Features of the summit under investigation here</td></tr>
</table>

Table 3.1: Defining and optional features of HIPS

The symbolic loading of HIPS may also be facilitated by contributions by civil society actors, either joining summit organizers in their preferred hymn celebrating the event or, as suggested by the view on media events as sites of struggle (see 2.3.2), exploiting them for their own counter-framing. The (5) *incorporation of civil society actors* can hence be argued to make a strong difference for how summits are constructed symbolically and to what degree they resonate with media and audiences. The COPs of the UNFCCC process have traditionally been very open to non-state actors (Oberthür et al., 2002).

Another characteristic making the type of summit under investigation here somewhat different from other HIPS is its (6) *annual recurrence*. From the perspective of a transnationalization of national public spheres (see 2.3.1), recurring summits are particularly interesting, since their recurrence could be theorized to be the backbone of gradual transnationalization processes. Especially in the context of long-term negotiations in the UN framework, the effect on national public discourses could be assumed to be a matter of years. "UN global conferences must be seen in a long-term continuum; the closing of a conference may very well be a pause in an ongoing process" (Fomerand, 1996, p. 372). In this way, they are dissimilar to once-off summits. To conclude, this study is hence about a particular type of HIPS – namely those providing space for symbolic resources, integrating actors from civil society, and taking place regularly (see table 3.1). The findings presented later pertain to summits with these features; they should not be regarded as widely applicable to all other types of HIPS. The features of symbolic loading and civil society involvement are returned to below.

3.2.1 A closer look: History and functions of summits

The rise of the summit as an instrument of diplomacy can be divided into two phases (Melissen, 2003, pp. 13-14): As already mentioned above, it firstly emerged around the middle of the last century as a *bilateral* (or trilateral) forum occupied with "the great problems of war and peace and the reshaping of alliances in an emerging bipolar international system" (p. 10). Facilitated by the spread of convenient jet travel (Dunn, 2004, p. 139; Melissen, 2003, p. 9), international politicians could now use their qualification as "people readers rather than paper readers" (Melissen, 2003, p. 2) and meet their counterparts face-to-face, often associated with high public expectations. Diverse functions are attributed to bilateral summits – some of more psychological value, such as the establishment of trust or the elimination of suspicion among politicians, others truly political in nature, like the gathering of information or the speeding up of policy processes (Dunn, 2004, pp. 150-153; Melissen, 2003, pp. 3-4). Common criticism of these meetings concerns their allegedly low effectiveness and often non-binding character. Accordingly, lack of substantive progress is often hidden under vague language in communiqués, "with a degree of ambiguity so as to leave room for manoeuvre for follow-up talks or the leaders' post-summit confrontation with their domestic constituency" (Melissen, 2003, p. 3). Whether such events actually have political effects "rests entirely on the political will and readiness of states to implement that to which they have agreed" (Fomerand, 1996, p. 365).

Supplementing bilateral summits, a new era of *multilateral* summitry emerged in the 1970s and intensified in the 1980s (Melissen, 2003, p. 14). This came in response to the growing perception that some of the world's most pressing issues could only be solved in teamwork (Fomerand, 1996, p. 373); "the reality of the growing interdependence of nations and the impact of globalization has made closer international cooperation and the development of appropriate institutions a necessity" (Weilemann, 2000, p. 19). Such institutions were partially created under the roof of the UN, but also as separate IGOs, like the numerous fora of 'regional diplomacy' (Dunn, 2004, pp. 144-145; Melissen, 2003, pp. 11-12), e.g. the EU, NAFTA, Mercosur, or ASEAN. For the field of environmental politics, Seyfang (2003) regards the UN Conference on the Human Environment, held 1972 in Stockholm, as the first multilateral summit acknowledging the need for an internationally concerted response to environmental problems. It was continued 1992 in Rio de Janeiro (the 'Earth Summit'), 2002 in Johannesburg ('Rio+10'), and 2012 once again in Rio ('Rio+20'). In assigned policy fields, UN global conferences allow for monitoring and information-gathering and "a process of action-oriented reflection and research that feeds and sustains international discussion" (Fomerand, 1996, p. 367). They may also be a platform for agenda-building. "If the political setting is favorable and auspicious – that is, if there is a sufficiently strong and influential web of political alliances – some questions will enter the agenda of the international community" (p. 371). Provided that a given summit is equipped with an appropriate mandate, these alliances may even shape international law. Large parts of domestic interest aggregation and policy-making processes are interlinked with such intergovernmental or supranational processes at multilateral summits, turning them into a "frequent and routine instrument of international diplomacy" (Dunn, 2004, p. 165).

3.2.2 *Provision of symbolic resources*

Regardless of their exact functions and substantive policy output, what is widely regarded as an omnipresent feature of summits is some element of symbolism or ritualism. They "usually involve a ceremonial dimension" (Dunn, 2004, p. 149), which often even "triumphs over substance" (Melissen, 2003, p. 18). And despite their feel as "staged events" (p. 13), as "ritualistic affair" (p.18), the full degree of staging is commonly invisible to the casual observer, as a Dutch diplomat recalls: "What remains concealed are the text writers, the endless rehearsals and all the other preparatory work without which that one performance would not take place" (Melissen, 2003, p. 7). Accordingly, the symbolic value may also be strategically built up by organizers for political purposes. Being shown on television in the company of other statesmen may have positive effects on the perception with domestic voters or international audiences (Dunn, 2004, p. 152). This turns summits' symbolic value into "propaganda value" (Dunn, 2004, p. 151; Melissen, 2003, p. 13).

Symbolic loading of 'peace ceremonies'

For the classic Cold War summits between US and Soviet leaders, Hallin and Mancini (1992) detect both above-mentioned mechanisms: "the Durkheimian sense of a shared sense of membership and the Habermasian sense of participation in dialogue" (p. 136). Drawing on Dayan and Katz's original conception of media events as "planned, symbolic performances staged for a media audience"

(Hallin and Mancini, 1992, p. 121), they see such meetings as facilitators of (partial) global integration. Firstly, transnational communication flows are strengthened, turning the summit into "an extraordinary opportunity for the states involved to override normal limitations on communication, taking the spotlight worldwide" (p. 125). Secondly, the actors involved, particularly journalists, resort to grand imagery – "a handshake across hostile worlds" (p. 136), political "magic" (p. 133) –, thereby highlighting the event's worldwide significance, if not even evoking "a global sense of community" (p. 127). Simultaneously, the ritualistic honoring of the event allows for occasional episodes of humanization, for emphasizing the summit's 'human side' (p. 129). In a post-Cold War world, however, the symbolic resources that once stemmed from the confrontation of superpowers are no longer pertinent; US-Russian summits are now "treated like ordinary political news" (p. 136).

Fluctuations in the degree of performance and ceremony surrounding summits could also be found in the context of a different political confrontation. Liebes and Katz (1997) scrutinize the media event character of several "peace ceremonies" (p. 255) between Israelis and Palestinians in the early 1990s (not all of them were summits in the strict sense outlined above, as the element of topically focused negotiation was missing). The authors point to a "contract among three partners – principals, broadcasters, and public – each of whom must affirm that the event 'deserves' media-event treatment" (p. 238). If such contract does not come into being, the event will lack symbolic resources and the media – having to choose between "reverential closeness" (p. 255) or "critical distance (ibid.) – will go for the latter. This is, however, no unilateral decision on the side of the media. Political organizers must agree on the summit's wider significance and behave accordingly in their communication, and the audience, for its part, must be ready to honor the occasion in a ritualistic manner (which would not be the case if, e.g., central summit figure were widely disliked). Even in the context of symbolically laden confrontation (US/USSR, Israelis/Palestinians), bilateral summits often lacked transnationalizing potential.

Symbolic loading of 'world conferences'

While the symbolic value of the traditional bilateral summit commonly stemmed from contexts of confrontation (as laid out above), the emergence of global governance as well as the dissolution of bipolarity in the international system have removed that frame of reference and freed summits of "ideological dead weight" (Weilemann, 2000, p. 19). However, this "diminished propaganda value of summitry" (Melissen, 2003, p. 19) in post-Cold War times should not be interpreted as a universal lack of symbolic resources attached to contemporary HIPS.

Most of today's multilateral summits even have "more formal character, with carefully orchestrated closing ceremonies and a considerable degree of pomp and circumstance" (Melissen, 2003, p. 15). This "element of theatrics" (p. 372), which Fomerand (1996) attributes to most UN summits, can be an effective carrier or facilitator of symbolic resources that various actors want to attach to the meeting. Besides their presentation as 'last chances', 'great opportunities', or 'likely failures', UN-run HIPS may stand for specific ideas or targets, possibly even lending "a 'halo of respectability' to ideas hitherto considered unacceptable" (Fomerand, 1996, p. 370). The ceremonial, ritualistic dimension of HIPS – detectible also in the sea of flag poles outside convention centers, the arrival of heavily protected motorcades, or maybe even the confrontation of protesters and riot police in the streets – serves as carrier or 'wrapping' for the aspirations and anxieties connected to these events. It may be this "ceremonial aspect rather than the substance of the negotiation or exchange of views that makes the multilateral summit 'real' to the general public" (Melissen, 2003, p. 16).

Making use of the notion of 'symbolic politics', Wesel (2004) argues that HIPS organized under the auspices of the UN constitute ideal platforms for political rituals and ritualistic media treatment. Being part of the UN framework bestows upon these gatherings an "aura of sacred, elevated significance" (p. 130, own translation) and unlocks a world of symbolic resources that political actors and media can exploit in their summit dramaturgy. The two letters 'UN' grant worldwide authority, and so do the organization's manifold symbols, such as the UN flag – maybe the first political symbol in human history that all people of the world accept (p. 143) –, blue berets, the architecture of UN headquarters, or the office of the UN Secretary General, who personifies some kind of "secular pope" (p. 130, own translation). The presentation and celebration of these and other global symbols at HIPS – Wesel calls them 'world conferences' (own translation) – bring about particular foci of ritual communication, such as establishing a (global) community (p. 238) or coping with (global) problems (pp. 241-243). Wesel argues that it would be wrong to perceive such performances of symbolic resources as mere facade and useless for political decision-making. They may actually be important facilitators of political progress (p. 229).

It becomes clear from these accounts that the symbolic loading of summits is commonly seen as collaborative effort by various actors. As mentioned above, media's 'performance' surely plays a large role in this, as the abandonment of normal reporting routines and the celebration of symbolic resources might emphasize the event's symbolic significance vis-à-vis audiences around the world (see 2.3.2). But, as Hallin and Mancini (1992) showed, media performance is dependent on contributions by other summit stakeholders. Widely noticeable involvement by various actors elevates the event's perceived symbolic significance and, thus, facilitates media performance. Within this process, visible resonance with actors from civil society, namely NGOs, may play a key role.

3.2.3 Participation of civil society

The exact nature of NGO participation at HIPS is heavily influenced by the regulations that principal organizers set for such involvement. From its inception, the UNFCCC process, whose summits are under investigation here, has been relatively open to NGOs; "the regime, in order to encourage NGO participation, has decided to employ a 'light touch' in managing their involvement" (Oberthür et al., 2002, p. 124). This is first and foremost visible in NGOs' opportunity of being accredited as official observer organizations to the Convention. In order to attain this status, organizations have to fulfill some broad criteria, such as being "qualified in matters covered by the Convention" (UNFCCC, 2010, p. 1) or serving non-profit purposes (ibid.). These conditions provide a helpful basis for deriving a definition of NGOs for this study (see also Betsill and Corell, 2008a, p. 4; Oberthür et al., 2002, p. 126-128). With some minor specifications of and additions to the UNFCCC criteria, I define NGOs as

organizations that (1) come into being and operate independently from (inter-) state entities, (2) possess non-commercial interests and/or expertise regarding political decision-making, and (3) aim at attaining their objectives in a non-violent manner by exercising influence on established (inter-)state procedures.

Being accredited allows NGOs to access the summit venue, take part in selected meetings by making oral interventions, circulate printed materials, and engage in lobbying (Oberthür et al., 2002, pp. 128-134). Especially the right to approach negotiators is often used. "NGOs spend considerable time trying to influence talks by persuading government representatives, who have the formal power to make the decisions, to accept the non-state actors' perspectives" (Corell and Betsill, 2008, p. 23). It should not be assumed, however, that interactions with NGOs are generally unwanted by governments. Especially larger NGOs from industrialized countries are treated as sources of valuable expertise by negotiators and are often consulted on the more intricate matters of the negotiations (Oberthür et al., 2002, p. 123). "The presence of NGO experts at the sessions of Convention bodies enables them to provide 'real time' analysis and advice to delegates" (ibid). It helps negotiators in "understanding the nature of the problems and the implications of various policy alternatives under consideration" (Corell and Betsill, 2008, p. 23). The mutual dependency of governmental and non-governmental actors can bring about stable relationships between summit participants on both sides featuring frequent informal exchanges through cafeteria meetings, phone calls etc. (Betsill, 2008a, p. 48). Also, it is common practice for NGO experts to become officially embedded in government delegations (Oberthür et al., 2002, pp. 134-135; Wesel, 2004, p. 193).

In general, the openness of the multilateral climate change talks to civil society as well as the latter's growing interest in this global issue has turned NGOs into an ever-growing, multifaceted element of these HIPS. "NGOs in the climate negotiations have moved well beyond a single profile and contribute to the process on many levels and with many voices" (Carpenter, 2001, p. 321). The largest NGO delegations at climate summits are traditionally sent by Greenpeace, Friends of the Earth, and WWF (Betsill, 2008a, p. 46). But also outside environmental politics, the presence of civil society at HIPS has expanded (Carpenter, 2001, p. 319; Fomerand, 1996, p. 363). This indicates a growing appreciation and respect by politicians for civil society concerns (Melissen, 2003, p. 8). The summit is thus "increasingly a place where diplomacy at the highest level meets public concerns, where political leaders will have to show their commitment to working with non-governmental stakeholders" (p. 21).

3.3 Public diplomacy at HIPS

Having defined and explored the concepts of public diplomacy (3.1) and HIPS (3.2), it is now time to bring both together. I first review the sparse literature explicitly dealing with efforts of strategic communication at such events (3.3.1) and then develop a conceptual model guiding my analysis of the communication strategies that governments and NGOs carry out on site (3.3.2).

3.3.1 Previous research

Research on how and why political actors use summits as locus of strategic communication is limited. Especially regarding *governments*, few scholarly efforts have been taken to theoretically conceptualize and empirically investigate their communication behavior at summits. While we may find statements that summits "may of course have electoral value" for political leaders (Melissen, 2003, p. 16) or that their staffs can usually be found carefully "considering, planning, and managing every symbolic message" (Schill, 2009, p. 24) stemming from their principals' participation, full-fledged studies on the topic are rare.

Within Gilboa's (2001) above-mentioned typology of possible constellations between political actors and media in international relations (see 3.1.2), media events, such as summits, show up in two contexts. Firstly, under the heading of 'media diplomacy', he describes them as communicative catalysts carefully crafted by state actors to mobilize public support for conflict resolution or policy change. This seems to apply mostly to the bilateral peace-making summits mentioned above. Secondly, referring to non-state, transnational public diplomacy, he describes them as communicative vehicles exploited by non-state actors for their communication objectives. What he seemed to have in mind here are cross-border discourses or happenings that are taken advantage of by activists. Clearly, summits may assume both roles at the same time, being crafted by one set of actors and simultaneously being exploited by others. Despite this conceptual vagueness, Gilboa's achievement lies in building a first conceptual link between public diplomacy and media events.

Governments' strategic summit communication

Hahn et al. (2008) provide a more comprehensive study of state actors' purposeful staging of political events for strategic communication purposes. Focusing on the six-month period in 2007 when the German government presided over the EU Council, they are able to show that government-organized events lead to high prominence of government actors and frames in public discourse. These effects could particularly be found for the top-level Berlin summit in late March 2007, which brought together the EU's heads of state and/or government for the celebration of the Union's 50th anniversary and passing of the so-called 'Berlin Declaration'. According to the authors, the event was meticulously designed to "incarnate the positive spirit of the Berlin Declaration and its value on the road to a Constitutional Treaty" (p. 344). Considering that the German government could successfully align EU coverage to its agenda and spread desired frames by means of this event, they conclude that "political events, and especially staged media events, aggregate and culminate different issues and related opinions, and can therefore be used by policy makers and spin doctors" (p. 346). While media fallout was only monitored for German outlets in this study, it does highlight the value of summits in achieving *primary*, i.e. summit-related, communication objectives.

Other contributions focus on the *secondary* effects that may result from a government's hosting of a (political) event. These are not about the short-term spread of particular messages or frames facilitating the attainment of political objectives but rather about more long-term image effects that the host country or city could profit from. Rivenburgh (2010) discusses the hosting of global media events like summits or world championships as part of governments' image strategies to "attract favourable, and saturated, media and public attention" (p. 187). The long-term goal behind such efforts may be the increase of visibility, buildup of prestige, or promotion of tourism and foreign investment (Giffard and Rivenburgh, 2000, pp. 8-9). Emphasizing the high risks associated with the hosting of such events, Rivenburgh (2010) contends that they can be regarded as an effective tool for image-building public diplomacy if (1) their thematic core is related to the existing image of the host country, (2) event operations run smoothly and hassle-free for the media, and (3) the hosting government does not already enjoy high negative-valence visibility in news coverage.

NGOs' summit communication inside summit venues

Literature is more comprehensive when it comes to *non-governmental actors*' utilization of summits. One of the most systematic contributions is Betsill and Corell's (2008b) investigation of the influence of accredited NGOs at environmental negotiation summits. Rooted in a political science perspective, they trace how on-site efforts by "NGO diplomats" (Betsill and Corell, 2008a, p. 3) bring about changes in the negotiation process and outcome that would not have occurred without NGO activity (Corell and Betsill, 2008, p. 24; see Clark, Friedman, and Hochstetler, 1998 for a similar approach)[10]. While their focus is on political decision-making processes rather than the media coverage or public image coming out of these summits, they see strategic communication of public nature as one way of attaining political results. This can happen, for example, through NGOs' resorting to "a strategy of 'blaming and shaming'" (p. 23), i.e. drawing attention to governments' "actions that interfere with the negotiations and/or noncompliance with previous commitments" (ibid.). For the case of the climate change talks, Betsill (2008a) highlights the NGO community's regular hosting of public side-events and publishing of briefing papers and bulletins, like the *ECO* newsletter: "a useful way for delegates to keep up with the day-to-day progress of the talks" (p. 47) and "in terms of exerting influence, […] a political forum for promoting [NGOs'] positions on a variety of issues" (ibid.). A comparable platform can be found in the Fossil of the Day award, which the NGO alliance CAN International confers every summit day in a humorous public presentation upon those parties found acting most destructively in the negotiations. It is "CAN's way of singling out those delegations that have in some way undermined the environmental integrity of the negotiations" (Oberthür et al., 2002, p. 135). Such strategic communication activities carried out on site may comple-

10 Incidentally, Betsill and Corell (2008a) employ an analytical framework and methodology that are not dissimilar to this study.

ment and reinforce direct, non-public lobbying by NGOs and play a role in influencing the negotiations.

NGOs' integration of internal and external summit communication

While Betsill and Corell (2008b) focus on how NGOs may influence decision-making *within* the heavily guarded confines of a summit venue, we should remind ourselves that a summit's symbolic significance may depend on its wider reception across all types of stakeholders (see 3.2.2), including the occurrences *outside* convention centers, in the host city's streets and squares as well as on newspaper pages, TV shows, and websites. McCurdy (2008) advocates a holistic approach to analyzing summits that considers both their 'representational' and 'immediate' elements (p. 295). While the former is constituted by how the event is covered in the media, the expectations raised and symbolic meanings attributed, the latter refers the various physical spaces that are connected to the summit. Importantly, both elements intertwine in the course of a summit, forming a "hybrid site" (ibid.), at which "dispersed places become interwoven through media narrative" (p. 296). These places are not only located at the official summit venue itself, but wherever media capture stories that are integrated into the summit narrative: in the lunch room of riot police officers, in the car of a local commuter stuck in a summit-induced traffic jam, or at an NGO panel discussion on the summit's likely failure. Especially civil society reactions have become a strongly anticipated component of media's summit performance (p. 295); HIPS become symbolically loaded only through "large and vociferous sideshows" (Giffard and Rivenburgh, 2000, p. 12) featuring a plethora of actors running diverse summit-related activities at various locations.

NGOs can hence engage in summit communication on the *inside*, i.e. as accredited observers allowed to enter the summit grounds, or on the *outside*, making use of infrastructure and public space beyond the confines of the venue (Beyeler and Kriesi, 2005, p. 107). While both settings provide room for campaigning, particular circumstances (such as security regulations) and exercisable activities might differ. As described above, on-site campaigning may include direct interactions with negotiators, whereas off-site campaigning is aimed at creating a particular tone of coverage or influencing public opinion and often comes in the form of (sometimes violent) protest marches. Both approaches can be successfully integrated, as Beyeler and Kriesi (2005) point out on the basis of an analysis of the protests at several WEF and WTO summits between 1994 and 2004. In particular, civil society actors' protest may add gravity to the positions advocated inside the venue vis-à-vis negotiators (p. 107).

NGO protests and protest camps at the G8 and WTO summits

Overall, the 1999 WTO conference in Seattle must be regarded as the "birth date" (p. 101) of widely noted protests attached to summits, at least for the counter-globalization movement. Media interest in the summit was already high in the first place – visible, e.g., in frequent declarations of the event's symbolic significance as the launch of the 'millennium round' of global trade talks – and then led to strong media attention to the clashes between protesters and police termed 'the Battle of Seattle' (p. 102). From this, Beyeler and Kriesi (2005) conclude that a "strategy of staging protests at summits only seems to help gain widespread media access when the protest target itself becomes a top media event" (ibid.). The value of summits as resources for activists thus depends on the presence of a "window of opportunity" (p. 107) granting them access to the mass media forum. However, such circumstances of heightened media attention might also result from journalists' anticipation of (violent) protest (McCurdy, 2008, p. 295), demonstrating how civil society actors and the media might mutually reinforce each other in raising the symbolic significance of the event. Through this mechanism, highly visible demonstrations have become a common component of summits, which are often not remembered "as quiet and civilized international meetings, but as beacons for loud and violent international protest" (Melissen, 2003, p. 20). This illustrates their character as "sites of struggle" (McCurdy, 2008, p. 294), where "conflicts of interest between governments and civil society will continue to be articulated" (Melissen, 2003, p. 20).

In addition to occasional protest, summits are often accompanied by permanent civil society platforms (Fomerand, 1996, p. 363). These 'protest camps', 'counter-summits', or 'alternative summits' are established by social movement organizations, citizen initiatives, or other NGOs as deliberate counterweights to the official event. Their external function lies in challenging the official summit's monopoly on media attention by counterbalancing it with an equivalent rooted in civil society. Internally, these platforms facilitate the networking and coordination between NGOs and provide space for critical analysis and discussion of what happens at the negotiations. In rare cases, even government representatives may pay a visit to such happenings (Beyeler and Kriesi, 2005, p. 105), although they are predominantly the domain of non-state actors and oppositional politicians.

Padovani (2010) explored the strategic communication practices of local activists at the 2009 G8 summit in L'Aquila, Italy, which were coordinated from a basecamp at a local park that also housed a so-called 'Media Lab' providing Internet access and media production facilities. Activists intended to use the "golden opportunity provided by the media exposure during the international gathering" (p. 419) to "perforate the government's media spectacle" (p. 426). Most successful was a stunt on the politicians' arrival day, when activists inscribed an oversized 'Yes We Camp' to a mountainside. Mocking Barack Obama's famous campaign mantra, the slogan was to draw attention to the fact that many of L'Aquila's inhabitants were still living in provisional housing after a recent earthquake. The giant letters could not be overlooked when entering or leaving the summit venue and made their way into international outlets (p. 431). Attention was also given to a march of the town's 'Last Ladies', who – in contrast to the statesmen's First Ladies – lived under miserable conditions and partly only wore underwear at the demonstration (ibid.). Both strategic communication activities were part of inventively themed, to-the-point campaigning exploiting the international event for promoting local causes. Especially the creativity put into devising photo opportunities and protest marches resonated well with the media (p. 436).

Studies on NGO campaigning at the 2005 G8 summit in Gleneagles, Scotland come from McCurdy (2008) and Sireau (2009). The latter explored the production structures and processes of the Make Poverty History (MPH) campaign that utilized the summit as climax and resorted to tools of advertising and branding for mobilizing against global poverty. The effort was run by a coalition of up to 540 member groups as well as academics and pop stars (p. 4). Few days before the summit kicked off in Gleneagles, MPH hosted a rally in nearby Edinburgh, featuring several celebrities and drawing 225,000 participants (p. 23). Worldwide attention to the G8 summit was also facilitated through the ten Live8 concerts that took place on the same day in all G8 countries plus South Africa. Allegedly, three billion TV viewers followed this globe-stretching event featuring more than 1,000 artists performing at all venues (p. 22). Finally, July 6, the opening day of the summit, saw a Live8 concert taking place in Edinburgh – the closing effort at mounting public pressure on G8 leaders to take decisions towards fairer trade relations, debt relief, and foreign aid improvement (pp. 4-5). Although Live8 and MPH were separate initiatives, they coordinated their activities (p. 187) and eventually ran the risk of being regarded as one and the same thing (p. 189).

While the MPH campaign clearly profited from worldwide attention sparked by Live8, some of its organizers worried that it was outshone by the concerts, which lacked clear messaging, or even contradicted by ill-briefed musicians (pp. 189-194). This concern proved justified when at the end of the summit, in its press relations, MPH criticized the summit communiqué for its lack of commitment (p. 24), while musician Bob Geldorf, founding father of Live8 and commonly seen as representative of MPH (p. 193), praised the document as containing all the steps demanded (p. 195). This essentially confused audiences;

"more negative assessments from NGOs discouraged some, while the overly positive ones left them perplexed" (p. 202). The episode points to the value of coherence and coordination within NGO communities. According to Sireau (2009), several internal conflicts – also, for instance, between adherents of brand-based marketing and message-based campaigning – characterized the production of strategic summit communication, turning it into "a process of negotiation between competing groups and factions" (p. 200).

One faction only scarcely represented in the MPH campaign was that of radical, anti-capitalist NGOs. A prominent member of this group was the *Dissent!* network (p. 25), which generally opposed MPH, as the campaign "had co-opted G8 protests in order to lobby the G8 instead of question its legitimacy" (McCurdy, 2008, p. 303). *Dissent!* set up a protest camp in the summit venue's vicinity, which was accessible to mainstream media at selected times only (pp. 299-300). One function of the camp was to serve as meeting point for (partially militant) activists engaging in blockades of roads to Gleneagles on the first summit day (pp. 300-302). While the goal to "shut down the G8" (p. 301) was certainly not accomplished, it should also not be understood literally, as McCurdy argues: "protest actions are not undertaken to achieve their declared aim […], but to create the appearance of attempting to do so" (p. 305). Blockades, 'shut-downs', and other forms of NGOs' direct action should hence be regarded as 'spectacular action' consciously and unconsciously planned for the media. Their purpose is "to create the perception of action; to provide the appearance of resistance. It is a ritualized performance of resistance placing emphasis on symbolic over physical disruption" (ibid.). Consequently, even though *Dissent!* refrained from participating in the government-friendly media spectacle of Live8 and MPH, it was still fulfilling its "allocated role" (p. 307) in the construction of a media event.

Empirical research on strategic summit communication: A colorful picture

This review of the few studies that explicitly treat summits as loci of strategic communication has created a colorful picture: While political actors surely regard and utilize these events as communication resource, motivations and strategies are diverse and cannot be easily systematized. On the side of governments, objectives may range from the promotion of particular issues and frames (Hahn et al., 2008) to the mobilization of public support for significant policy changes (Gilboa, 2001), while secondary image effects might also be desired (Giffard and Rivenburgh, 2000; Rivenburgh, 2010). Specific summit-based communication activities by state actors remain under-researched, however. On the side of NGOs, we have to distinguish between on-site lobbying supported by targeted public pressure (Betsill and Corell, 2008b) and large-scale external protests (Beyeler and Kriesi, 2005), which are often coordinated in protest camps (McCurdy, 2008; Padovani, 2010) or NGO coalitions (Sireau, 2009). NGO campaigning may draw effectiveness from creativity (Padovani, 2010), but should

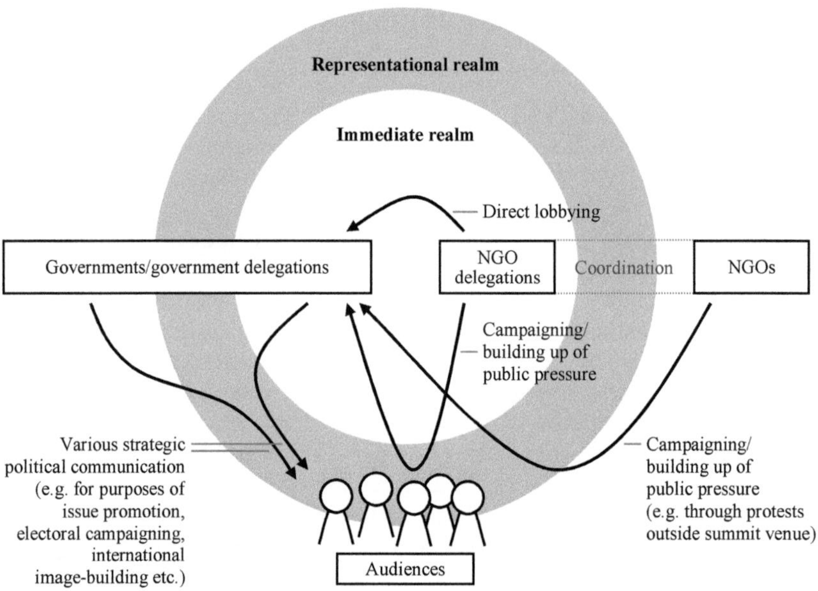

Figure 3.3: Routes of strategic communication at summits

avoid messaging inconsistencies and internal quarrels (Sireau, 2009). Overall, strategic communication fuses summits' immediate and representational realms, as expectations and meaning attributed to the event by the media may also shape physical events, like protest (McCurdy, 2008). An attempt at visualizing all described routes of strategic summit communication can be found in figure 3.3.

3.3.2 Conceptual model of the development of communication strategies

Based on the reviewed literature, I now outline a conceptual model that can explain how political actors' public diplomacy at HIPS is shaped, i.e. which factors influence their choice of communication strategy. This deductively designed model serves several purposes: Firstly, it ensures the theoretical grounding of my research, as it focuses the empirical work on aspects that have already been made subject to some theorizing (even if that is of rather preliminary nature in some cases). This way, my findings can be better integrated with the state of research and contribute to a true accumulation of knowledge (see 4.1.4 on this point). Similarly, the conceptual model constitutes what in other studies are the hypotheses. It contains assumptions on which aspects of summit business may be linked with others, where causal links, or at least relationships of influence, may be found. These assumptions are surely no hypotheses in the statistical sense, which posit quantifiable degrees of association between variables. Instead, they outline (based on existing literature) likely relations, interactions, and processes at HIPS. This caters to the approach of process-tracing, which – together with structured, focused comparison – constitutes the key methodological concept drawn on in this case study (see 4.1.4). Finally, the model strongly structures my case study. The various relations and influences proposed in the model are examined bit by bit in the empirical part and the different theoretically conceived influences checked against empirical reality (see section 5.1). The findings of this empirical validation are then worked into a revised version of the model (see figure 8.1).

As laid out in sections 3.2 and 3.3, HIPS are understood as public diplomacy resources that political actors may use for efficiently influencing public discourse around the world to their advantage. HIPS are episodes of high political stakes, communicative effort, and organizational investment. This also entails increased levels of mutual observation among actors and their readiness to reflect on, and possibly readjust, their public standing at these events. Actors may routinely evaluate the effectiveness of their own and other actors' communication strategies and consider such assessment in their strategy building. According to this perspective, *continual readjustment to changing circumstances* can be seen as integral part of strategic communication. To honor this, my model features two fundamental feedback cycles – one short-term, one long-term – that link the outcome of a given strategic communication process to future processes. The various components of the model, which is outlined in the following, are lined up along these two feedback cycles.

At the heart of the model (see figure 3.4) lies the explanandum of a given political actors' *communication strategy* and corresponding *communication activities*. The distinction between 'strategy' and 'activity' is inspired by the arena model of the public sphere (see 2.1.1), which conceives of public speakers as being active on a 'back-stage', where they read the communicative terrain lying in front of them and craft their contributions to public discourse, and in the 'arena' (or on the 'front-stage'), where they publicly present these contributions. An actors' strategy is crafted on the backstage; it establishes a communication objective (e.g. raising awareness for an issue or increasing public pressure on an actor) and the sequence of communication activities to be carried out for attaining that objective. Hence, activities are publicly visible acts of communication (e.g. at media briefings or photo ops) that are prescribed by strategies. This sequence of developing strategy and implementing corresponding activities can also be referred to as 'strategic communication process'. It is built on the general *communication structure and processes* that a given political actor operates on the occasion of a high-level international political summit. While 'structure' refers to more stable organizational elements – such as dedicated communication teams, material and non-material resources, or hierarchies and reporting lines –, 'processes' refers to sequences of human action that are based on and limited by these organizational elements (Giddens, 2001, pp. 667-669).

Structural/long-term factors shaping political actors' communication strategies

As part of actors' communication structures and processes, strategic communication processes at HIPS may be shaped by a variety of more long-term factors, which are either decided upon prior to the summit or even inherently linked to the nature of the organization.

One of such 'inbuilt' factors that may shape a political actor's communication strategy is the *type of actor*, i.e. whether it is of state (governmental) or non-state (non-governmental) nature. At HIPS, governments play a thoroughly different role from NGOs. As 'parties' to the UNFCCC, they are legally embedded in negotiations and exercise political influence within decision-making procedures. While this may be supplemented by the exercising of soft power through public diplomacy (see 3.1.1), governments' prime orientation is to engage in policy-making. This is different for NGOs, whose influence on the negotiations rests almost entirely on soft power (see 3.1.3) and who may therefore resort to more elaborate communication strategies. Generally, governmental and non-governmental PR show clear differences (see, e.g., Pfetsch and Wehmeier, 2002): While governments often have dedicated agencies for promoting public acceptance through classic media relations and informational work, NGOs voice their interests vis-à-vis the state through campaigns whose unconventionality and creativity may compensate for limited PR resources.

Another rather stable factor that may shape the kind of communication strategy that political actors carry out at HIPS is their *worldviews and standpoints* in the negotiations and debate around climate change. Here, different aspects may play a role, such as the extent to which a government (i.e. country) is affected by climate change, which forms of mitigation and adaption it prefers out of economic reasons, or its integration into alliances. Hence, an AOSIS government facing the immediate threat of rising sea levels might use HIPS for different communication strategies than a newly industrializing economy profiting from weak transnational climate regimes. But also on the side of NGOs, contrasting approaches to combating climate change can be found, which may be visible in differences in the style of rhetoric or the degree of confrontation drawn on in communication strategies. Doyle (2009), for example, shows that ENGOs may fundamentally differ in the values they embody and that this has an effect on how they publicly frame the issue (see 3.1.3).

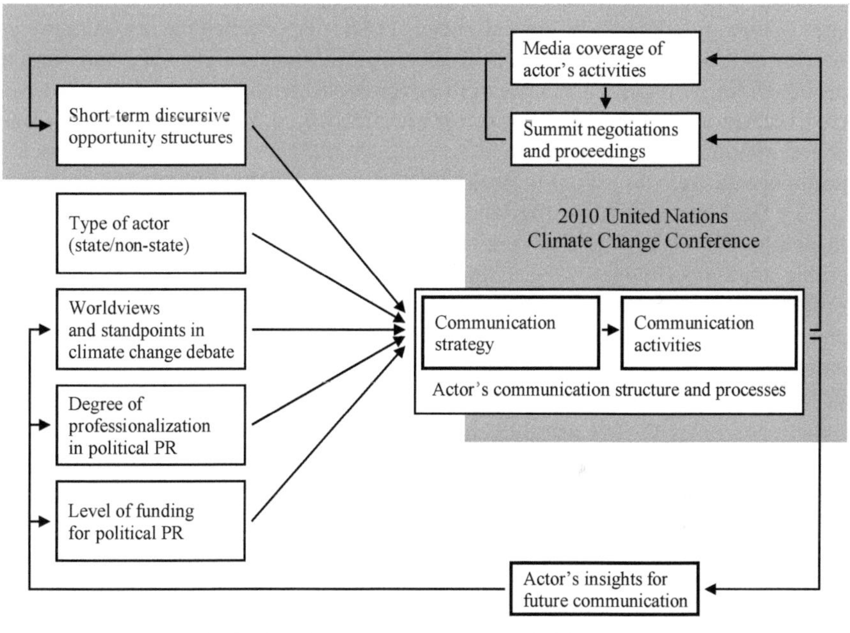

Figure 3.4: Deductively developed conceptual model (prior to empirical testing)

Differences in summit communication strategy may also be attributable to political actors' varying degrees of *professionalization* in their political PR. I follow Röttger, Preusse, and Schmitt (2011, pp. 249-260) in conceptualizing professionalization as the development of occupational fields into independent 'professions' that are based on dedicated expertise (often grounded in scientifically grounded knowledge) and particular privileges, such as relative autonomy in taking decisions and organizing routines. Professionalization is closely related –

but not equivalent – to 'professionalism', which rather denotes the quality of individuals' conduct and of products of their work. Accordingly, professionalization refers to overall developments that might facilitate professionalism. With regard to summit PR, the degree of an actor's professionalization might primarily be detected by analyzing communication structures and processes: To what extent are individuals tasked with communication functions true 'professionals', i.e. to what extent do they hold specific PR expertise and are acknowledged by their organizations as proficient engineers of communication strategy? Or to what extent do strategic communication processes incorporate dedicated PR expertise and are exclusively oriented towards the attainment of particular communication objectives, such as the influencing of public opinion or media coverage? While governments commonly have at their disposal dedicated communication capacities, their degree of professionalization may surely vary and should probably not be regarded as crucial to government' success. For NGOs, however, which to a much higher degree rely on effective communication, professionalized conduct with publics and the media should be seen as precondition for political achievement (Schmitt-Beck, 2001, p. 32; Sikkink, 2002, p. 315).

A final preset factor is the *level of funding* that political actors can draw on in summit PR. Using HIPS as communication resources surely requires a considerable amount of money – be it for communicators' airfare and hotel bills, the printing of information brochures, the renting and furnishing of booths, or the contracting of external consultants. This is not to say that ample funding does automatically translate into effective strategy (as mentioned above, NGOs often generate communicative efficacy *in spite of* or maybe even *facilitated by* limited funds), but we can still generally assume that more money also means more leeway in carrying out those communication strategies that are deemed effective by actors. While the work of government delegations at HIPS is typically financed by the respective ministries through state funds, NGOs' income is less steady, since they rely on individual donations or project-based resources granted by foundations and state agencies (Take, 2007, p. 199). Due to this volatility, financial capacity and its effects on strategic behavior can be assumed to be a more prominent topic on NGOs' internal agendas than on those of governments.

The specific manifestations of these four variables for a given actor are decided upon through principal organizational processes, such as budget planning or policy formulation, which take place outside summit periods at actors' headquarters (ministries, NGO secretariats etc.). They are hence of more stable nature and not changed ad hoc ('type of actor' even is an intrinsic factor that cannot be changed at all). It is assumed in this conceptual model that following HIPS, actors evaluate the particular turning out and suitability of their communication structures and (strategic communication) processes at the event and make adjustments to the three modifiable factors, if deemed necessary. The particular nature and thoroughness of these evaluations has to be explored empirically, but some self-observation and self-assessment can be taken for granted, especially for social movement organizations, which della Porta and Tarrow (2005) describe as "'reflective' actors" (p. 8) that continuously evaluate the success of their strategies and tactics. Such review generates *insights for future communication* and might also include lessons for a reconfiguration of the three modifiable factors (e.g., through an increase in budgets or a radicalization of positions). It thereby affects the parameters under which communication strategies at subsequent HIPS are conducted.

Dynamic/short-term factors shaping political actors' communication strategies

Besides this first feedback cycle, strategic communication processes at HIPS might also be subject to more short-term adjustments over the course of the ongoing summit. The need for making such adjustments might arise when an actor's communication strategy turns out to be unsuitable for the communicative terrain of the summit forum or when summit proceedings bring about occurrences with particular implications for communication strategy.

In particular, such developments may trigger *short-term discursive opportunity structures*. Following Koopmans and Statham (1999), the concept of discursive opportunity structure refers to those factors in a society's political culture that facilitate the resonance of particular positions and argumentations with the public, while putting others at disadvantage (p. 228). Discursive opportunity structures are gateways to public discourse that stand open for particular discursive strategies; an actor operating unsuited strategies "will find no support for its ideas and demands, nor will it be able to gain access to the polity" (p. 247). They are different from political, or institutional, opportunity structures (see, e.g., McAdam, 1996) in that they do not touch upon the institutional, legal provisions regulating admission to political processes (such as minority representation clauses or election thresholds) but instead focus on parameters of culture and public discourse. Applying the concept of discursive opportunity structures to HIPS, they may be regarded as distinctive circumstances under which particular contributions to summit discourse are induced and others are discouraged. They are of *short-term* nature, arising from actors' unforeseen negotiation moves or discourse contributions (as opposed to the more stable cultural values in national societies). Short-term discursive opportunity structures might pose threats or grant unexpected options to the preplanned communication strategies of political actors and hence cause according adjustments.

As ad-hoc products of summit business, of the interplay between negotiations and summit PR (see 3.3.1), the emergence of short-term discursive opportunity structures might as well be influenced by strategic communication programs that political actors carry out at HIPS. This is the second, summit-internal feedback cycle. Through summit PR, political actors might trigger other actors' moves. These moves bring about particular opportunities that again have to be considered by other actors and influence communication strategies. This highlights the character of HIPS as special episodes of mutual observation and continual repositioning. The density of interactions at these events accelerates strategic communication and supplements the long-term, structural aspects shaping these processes with short-term, interactive factors.

3.4 Summary of theoretical argument

This chapter – together with the preceding one – established the theoretical context for the following empirical investigation of political actors' strategic communication efforts at the 2010 Cancún climate summit. Such efforts aim at crafting (or 'engineering') public discourse, the publicly visible exchange of positions among speakers. Public discourse takes place in the public sphere, a network of various fora on different levels, which are connected through communicative linkages of varying strengths. The 'master fora' within national public spheres are constituted by national mass media; they selectively provide visibility to public discourse taking place in other fora – be they of topically specialized, subnational, or also transnational nature. One type of transnational fora are high-level international political summits (HIPS), which episodically interconnect with national forums and thereby facilitate short-term transnationalization. If summit-induced transnationalization processes affect multiple national public spheres concurrently, we may speak of the emergence of a transnational public sphere surrounding the summit. Such conceptualization of transnational public spheres as being formed by transnationalized national public spheres appears to be more theoretically complex and empirically justified than proposals of a distinct transnational public sphere 'floating' above and largely autonomous from national public spheres.

The transnationalizing capacity of summit discourse is facilitated by the fact that HIPS may be loaded with symbolic resources that are co-constructed by various summit stakeholders, particularly those civil society actors enjoying high degrees of access to UN climate summits. The 'performance' of these resources in summit coverage may introduce a ritualistic element into the summit's reception and thereby strengthen its impact around the world. The crafting of these resources and other components of summit communication is subject to political actors' public diplomacy. Their transnational reach turns HIPS into strategic resources that allow actors to distribute their messages around the word in an efficient manner; attaining comparable degrees of attention without the help of HIPS would be far more expensive.

While the state of research on governments' and NGOs' summit communication is limited, some factors that were found to affect the configuration and effectiveness of public diplomacy in these settings are: the degree of cultural compatibility between originator and addressee; the combination of national transnational routes of communication ('two-level gaming'); the orientation towards primary (summit-related) vis-à-vis secondary (image-related) messages; and the challenges posed by internal coordination within NGO alliances. All explored factors were synthesized in a conceptual model that aims at explaining political actors' choice of summit communication strategies. In the model, HIPS are conceptualized as special episodes of close mutual observation and continual repositioning. Summit communication strategies are shaped by both long-term/structural and short-term/dynamic factors, which may partially be adjusted in response to the specific unfolding of events at the summit.

4. Investigating strategic communication at summits

This chapter aims at establishing transparency with regard to how I went about answering the research questions formulated above (see section 1.2). The variety of issues touched upon here ranges from more abstract methodological questions (like how to conduct case studies) to very concrete methodical, and even logistical, points (such as how I recruited my interviewees or ensured the confidentiality of sensitive information). What unites these discussions is the objective of making my empirical decisions open to scrutiny, of making evident that the findings presented in subsequent chapters were obtained by means of a well-reflected and systematic process. The diligent employment of methods is one of the core characteristics of scientific work (Brosius, Koschel, and Haas, 2008, p. 25; Diekmann, 2007, p. 18), and this chapter is hence to demonstrate the scientific quality of the research at hand.

In its first section, this chapter features an introduction to the general characteristics of the case study approach and how I configured it in this particular research (section 4.1). The two subsequent sections then discuss case selection, or sampling (sections 4.2 and 4.3), which also involves questions of generalizability and comparability. Details of data collection are clarified in section 4.4, with most subsections devoted to the methodical and logistical aspects of doing semi-standardized interviews. Section 4.5 focuses on data analysis and presents the steps of my computer-aided qualitative content analysis. Due to their detailed technical nature, sections 4.4 and 4.5 are not included in this volume but can be downloaded from this book's companion website (part of the Springer VS online presence).

4.1 The case study approach: basic features and specific configuration

The findings of this research were obtained through a comprehensive on-site case study of the communication strategies of governmental and non-governmental delegations at the 2010 UN climate summit in Cancún, Mexico, based mainly on semi-standardized interviews with communication officials. While the methodology of case studies might carry the connotation of an unsystematic and less rigorous approach in the eyes of some scholars (for corresponding debates, see George and Bennett, 2005; Gerring, 2004; or King, Keohane, and Verba, 1994), the aim of this section is to make a case for the opposite: case studies may constitute a structured and analytical way of exploring social phenomena and internal causal mechanisms. As with any other methodology, their scientific value depends on how systematic and diligent they are conducted (Jahn, 2006, p. 320). If carried out in a disciplined manner, case studies can be regarded as highly valuable and complementary to other research designs.

One of the most detailed texts on how to do high-quality case studies in the social sciences comes from George and Bennett (2005). For them, a case study is "the detailed examination of an aspect of a historical episode to develop or test historical explanations that may be generalizable to other events" (p. 5). While the attribute 'historical' may indicate a focus on *past* events, where researchers have to rely on already existing sources (i.e. mainly by means of archival work), the authors' definition seems also applicable to *contemporary* events, which researchers can make subject to their own collection of primary data and do not bind them to the information gathered by others.

George and Bennett see the methodology's defining feature in a particular way of *analyzing* social reality, i.e. through the technique of process-tracing (see below), while for Gerring (2004), it is about *selecting* aspects for analysis, "a spatially bounded phenomenon [...] observed at a single point in time or over some delimited period of time" (p. 342). The actual analysis, he writes, is not necessarily different from 'cross-unit', or large-N, studies. Despite such differences, both accounts are united in the wealth of advice they provide for conducting case studies and can hence serve as the foundation of this section. Some fundamental aspects of doing case studies raised there are worth highlighting.

4.1.1 Causality in case study research

The main advantage of case studies – especially vis-à-vis quantitative approaches involving statistical analyses – is their attention to context and detail (George and Bennett, 2005, pp. 19-22). While case studies are still located within a positivist framework involving variables and causalities, they allow for more holistic measurement of concepts than by single indicators and, hence, ensure higher validity. Claims of causality are not based on a mere covariation of variables but can be substantiated through the careful identification of causal mechanisms, their internal sequences and connections. Aim of any case study researcher should be the uncovering of an "uninterrupted chain of evidence" (p. 21) or, in the words of Gerring (2004), "to peer into the box of causality to the intermediate causes lying between some cause and its purported effect" (p. 348). This rigorous analysis of intervening variables and unobtrusive links is also called 'process-tracing'. It helps researchers to avoid what is commonly called the 'degrees of freedom problem' (George and Bennett, 2005, pp. 28-30) – i.e. difficulties in interpreting empirical evidence in favor of only *one* out of several possible explanations –, as observations of variables are multitudinous and comparison against observable implications of competing theories is thus facilitated. I return to process-tracing below (see 4.1.4).

4.1.2 Properties and value of case study findings

A nuanced, qualitative treatment of variables has implications for the nature of findings that can be obtained (pp. 25-27). Case studies will rarely provide quantitative information answering questions like 'how much' or 'of what value', i.e. for causal weight or strength of effect. Instead, case studies aim at exploring 'whether' and 'how exactly' social phenomena take place. This focus on the circumstances and trajectories of events is a major advantage for theory development. Through case studies, new factors, variables, and hypotheses can be discovered – and, eventually, new theories developed. However, the findings of case studies can also be used for testing theories. They could either suggest the rejection of a theory, especially when the case study was based on a 'crucial case' (Eckstein, 1992), or a modification of its scope conditions, i.e. its factors of necessity and sufficiency.

A frequent critique of case studies deals with the issue of generalizability (George and Bennett, 2005, pp. 22-25, 30-32). According to these arguments, the small number of cases studied, along with their potentially biased selection, seriously reduces the extent to which findings can be regarded as representative for other cases. George and Bennett argue that case studies should actually *not* aim at achieving wider representativeness: "Case study researchers are more interested in finding the conditions under which specified outcomes occur, and the mechanisms through which they occur, rather than uncovering the frequency with which those conditions and their outcomes arise" (p. 31). Instead of producing more abstract but generalizable findings, case studies are better suited for providing context and explanatory richness. This, however, does not mean that case study findings cannot be applicable beyond the cases under investigation. In fact, if the selected cases can be argued to share key characteristics with a wider, clearly defined population, contingent generalizations are in order (despite some remaining risk of over-generalization).

Jahn (2006) draws attention to the fact that compared to other methodologies, case study research centers very much around the individual researcher (pp.

335-336). Her or his cultural, political, theoretical, methodological, and methodical predispositions can significantly influence the research process, from fundamental questions of design and methods to the analysis and contextualization of data. Wherever possible, these restrictions and biases should be kept at a distance and research tasks confronted with little prejudgment. An obvious way to do this is to reflect on one's individual predispositions and to ask oneself throughout the research how it could be affected by one's own personal outlook.

4.1.3 Nomenclature: 'classes of events', 'units', and 'cases'

Despite some clear differences in the reasoning behind analytical case studies on one hand and statistical large-N studies on the other, both share a basic outlook based on variables and causalities (at least in the view of George and Bennett, 2005). This implies that even in case studies, with all their explanatory richness and attention to detail, decisions have to be taken on which aspects, or variables, to focus on – and which to ignore (p. 67). Essentially, a case study focuses on "a well-defined aspect of a historical episode that the investigator selects for analysis, rather than a historical event itself" (p. 18). It is never a *complete* event that is studied but only selected features that are of particular (theory-driven) interest.

This well-defined aspect, or selected feature, is what George and Bennett call a "class of events" (p. 17) – nothing more than a "phenomenon of scientific interest" (ibid.), something that several events have in common and that is worth focusing on. A 'case', in turn, is one particular "instance of a class of events" (ibid.), one unique phenotype of the phenomenon under investigation, one distinct way of manifesting itself at a particular event. This labeling becomes clearer in the context of my study: The class of events is 'public diplomacy by political actors at HIPS', also called 'political actors' summit PR'. Their strategies form the overall object of the study, the phenomenon of general interest. The particular instance of this phenomenon (the case) is the 2010 UN Climate Change Conference in Cancún, Mexico. Essentially, this case is only interesting as a vehicle for summit PR – any other comparable event would have done the job (see section 4.2). For Jahn (2006, p. 324), this step of clarifying the exemplary aspects of a case and reflecting on the relation between both levels of a case study – the particular and the general (Gerring, 2004, p. 346) – is what characterizes analytic case studies.

The nomenclature by George and Bennett is refined further by Gerring (2004). He replaces the term 'case' with 'unit' – and reserves the former for entities below, or within, the instance under investigation. According to this labeling, case studies would actually have to be called *unit* studies – with a 'unit' being a "spatially bounded phenomenon [...] observed at a single point in time or over some delimited period of time" (p. 342). In the case of this study, that spatially bounded phenomenon would be the summit. Units, in turn, contain sub-elements that are the actual basis for data collection – the cases. In this research, the cases are the individual political actors planning and implementing communication strategies. Cases, then, can be observed on several dimensions, or variables. Taken together, these concepts resemble the different components of a dataset (p. 342): cases can be conceived as rows, dimensions could be the columns, and the unit would be the dataset as a whole. Figure 4.1 visualizes this nomenclature for the research at hand.

4.1.4 Integrating process-tracing and structured, focused comparison

Case studies can follow diverse motivations, which vary mostly according to their incorporation of theory and their generalizing ambitions. George and Bennett (2005, p. 213) and Jahn (2006, p. 325) distinguish between (1) atheoretical *configurative-ideographic* case studies, which describe the uniqueness of a case without being guided by theory or bringing about more general patterns or hypotheses, (2) hypothesis-generating (or theory-inspired) *heuristic* case studies identifying patterns and hypotheses as basis for theory development, (3) theory-based, interpretative *disciplined-configurative* case studies, which interpret a case through the lens of a particular theory, (4) *theory-confirming* or *theory-disproving* case studies, which test theories, usually based on 'crucial cases', (5) *plausibility probes*, which serve as pre-tests for more extensive testing or development of theories, and (6) *deviant* case studies, which explore why a particular case is not part of general patterns. These six strategies should not be seen as exhaustive or mutually exclusive, but they reflect the variety of research purposes that case studies can be used for.

Although no perfect fit, the case study presented here comes closest to type (2). It is heuristic in the sense that strategic communication at HIPS, especially the transnational element and event orientation, had not been researched in detail beforehand and that it attempts to come up with a way of modeling such activities, which could then lead to further theory development. This endeavor, however, did not take place without any theoretical grounding. Various streams of literature were used not only to provide the wider theoretical setting of the study and develop its conceptual model, but also to guide data collection in the field (interview guides were deduced from the model). Hence, the inductive aspects of the study, necessitated by its heuristic motives, were confined by some deductive boundaries. While this might reduce the likelihood of radically new 'outside-the-box' findings to be produced, it at the same time effectively integrates the study with related research and contributes to a true accumulation – vis-à-vis just an uncoordinated piling – of knowledge (George and Bennett, 2005, p. 70-71).

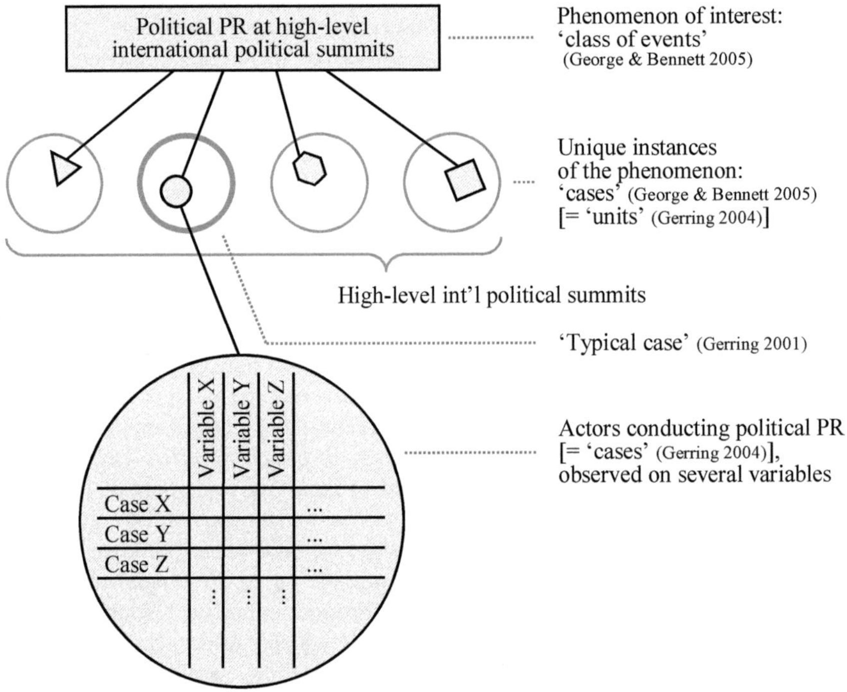

Figure 4.1: Methodological nomenclature used in this study

Guidance by theory is also needed for establishing the dimensions of comparison in a case study. Gerring (2004) argues that case studies always incorporate comparative thinking and are logically impossible without it (p. 342-344). This comparison can either work across different cases (to be understood as sub-elements of a particular unit, i.e. as the political actors at the Cancún summit), across different points in time (i.e. in the middle and after the summit), or in both ways. At least one of the two types of variation has to be present in order to infer some causality: "A single unit observed at a single point in time without the addition of within-unit cases offers no evidence whatsoever of a causal proposition" (p. 344).

The case study presented here put to use *both* kinds of variation. Firstly, it looked at several political actors – both governmental and non-governmental ones – and their communication strategies (see section 4.3 for the selection of political actors). The idea was to compare to what extent different factors influenced the development and implementation of strategies across these actors. Jahn (2006) points out that in small-N research, which is not able to statistically quantify causal weight, such comparison is the only way of identifying the credible ones among the many possible causal stories on offer (p. 321). Simultaneously, the case study compared the state of political actors' communication strategies over time, i.e. between the beginning/middle of the summit and its end/aftermath. The idea was to assess possible changes in communication strategies depending on the course of the summit and other actors' strategies. This way, short-term interactions between actors at the summit as well as more future-oriented lessons for the next summit or other communication activities could be determined. All in all, this approach resembled process-tracing, as it tracked down the decisions and interactions shaping particular outcomes (i.e. communication strategies) and thereby attempted to uncover causal sequences that would otherwise have been invisible.

In summary, the design of this case study effectively combined process-tracing with what George and Bennett (2005) call a 'structured, focused comparison' (p. 67-72): Each political actor was studied based on a standardized set of variables, ensuring comparability and accumulation of knowledge. At the same time, multiple points of observation allowed for the reconstruction of changes on these variables. The goal was to understand – and compare – the long-term factors shaping actors' communication strategies as well as the more short-term behind-the-scenes workings at such an event that also affect strategies. My conceptual model focused the comparison on selected aspects of the summit, but data collection still remained sufficiently open for findings outside these parameters. A deductive, theory-nourished perspective was hence combined with the more inductive approach of a heuristic case study suitable for theory development.

4.2 Unit selection: the Cancún summit and the climate change debate

When studying public diplomacy at HIPS – this research's phenomenon of interest –, one decision has to be taken fairly early in the research process: Which summit(s) should be in the focus? Which instance(s) of the phenomenon should be selected for investigation to learn about the wider 'class of events' (see 4.1.3)? This is a decision of sampling, of selecting from a wider population of possible cases the ones to be investigated. In this research, the population includes all HIPS in various policy fields that may be subject to symbolic loading and resonance from civil society (see section 3.2), such as the G8 or G20 summits in economic policy, the WTO meetings in trade policy, or widely-noted donor conferences in development and regional stability policy. Out of this multitude of cases – or *units*, drawing on Gerring (2004) –, I selected the 2010 UN Climate Change Conference in Cancún, Mexico, to serve as my object of investigation.

Climate change conferences (also called climate summits or COPs) have taken place annually since the establishment of the United Nations Framework Convention on Climate Change (UNFCCC) in 1992. The Cancún meeting was the 16th high-level summit within the UNFCCC process, which sets the frame for international political action against climate change. The Kyoto Protocol, a legally binding treaty built on the UNFCCC, was adopted in 1997 and came into force in 2005. The expiration date of the Protocol's commitment period was set to the end of 2012, and at the time of writing, countries were still negotiating a possible continuation of a global climate change regime in the post-Kyoto era. Expectations of a conclusion of a new treaty ran extremely high in the run-up to the 2009 climate summit in Copenhagen but were essentially disappointed. As part of a general post-Copenhagen fatigue, the 2010 Cancún summit was viewed with more skepticism and lower expectations.

It is my argument here that lowered expectations did not automatically impinge on the event's function as point of attachment for symbolic resources or as public diplomacy resource. As a matter of fact, in this atmosphere of climate action weariness, ENGOs and some governments with climate-friendly agendas felt that it was even more necessary to send out the message of climate change requiring an urgent response. The failure of COP-15 awarded to COP-16 the status of a "turning point" (Russel, 2010, ¶ 25): "Many climate experts, as well as journalists, question whether the cumbersome UN Framework Convention on Climate Change will even survive after this in its current form" (ibid.). Similarly, Boykoff (2012) pointed out that prior to the summit, observers wondered "whether negotiations in Mexico would solidify the backroom deliberation trends established in Copenhagen or whether the UN would boldly reassert itself and its relatively transparent, consensus-based process" (p. 251). While COP-15 certainly constituted a once-in-a-decade opportunity for political progress and became a 'mega summit' witnessing massive political PR investment, COP-16 should not automatically be considered as unsuitable object of investigation. After the disaster of Copenhagen, Cancún was to demonstrate whether the UNFCCC process was condemned to die or if some hope was justified. The summit mattered quite a bit and can hence be viewed as important PR platform.

A second argument for the selection of the Cancún summit could be formulated post hoc: As several interviewees confirmed to me, Cancún represented a 'typical' climate summit to a much greater extent than the rather unique Copenhagen meeting. Analyzing public diplomacy at Cancún can be assumed to reveal more of the *usual* communication processes at climate summits than an investigation of out-of-the-ordinary Copenhagen. In methodological terms, COP-16 can therefore be argued to come close to a 'typical case', which is defined as the "most usual case in a particular population" (Gerring, 2001, p. 218). While it might seem difficult to differentiate exactly between the usual and unusual parameters of a climate summit, the assessment by experienced summit participants that Cancún was much like an ordinary summit within the UNFCCC process provides good reasons for treating some of the findings as applicable beyond just COP-16. However, when it comes to making generalizations beyond the UNFCCC process, we have to be careful. A special feature of climate summits is their high degree of accessibility for civil society actors (see 3.2.3). This looks different for other international negotiation summits, where NGOs have to resort to other communication strategies than the ones found in Cancún.

Beyond these two arguments speaking in favor of selecting the 2010 UN climate summit as unit, the global debate surrounding climate change can generally be seen as highly suitable context for an investigation of processes of transnational communication. Climate change scores high among the most-widely noticed and hotly debated issues around the globe (Edwards, 2001; Miller, 2002; Olausson, 2009), and while media treatment of the issue and the climate summits in particular seems to differ from country to country (see 2.2.2.2), "it is *the* paradigmatic case to look for to encourage the emergence of transnational or global public spheres" (Eide and Kunelius, 2010, p. 12, italics theirs). To a higher extent than other issues of worldwide interest (such as terrorism or the financial crisis), climate change poses a threat to everyone on the planet, even though disparities with regard to responsibility and impact remain (Beck, 2007). The corresponding debate surrounding the effects of climate change and possible counteraction is therefore an ideal field for investigating research questions relating to transnational communication.

4.3 Case selection: governments and NGOs at the Cancún summit

After determining a case study's unit, the next step is to take a look inside that unit and select particular cases for analysis – those entities that serve as the actual basis for the approach of structured, focused comparison described above. This constitutes a second sampling decision, as it once again selects elements from a larger population. The population is composed of all governmental and non-governmental delegations (the ‘political actors’) present in Cancún and assumed to run strategic communication programs targeting audiences in more than one country. Out of this group, eleven delegations were chosen according to different criteria. From the governmental delegations, six countries with high visibility in the climate change debate were selected: the leading industrialized or newly industrialized democratic country from each continent with the exception of Australia and Antarctica (*Brazil* for South America, *Germany* for Europe, *India* for Asia, the *United States* for North America, and *South Africa* for Africa) as well as the host country *Mexico*. On the side of non-governmental delegations, the five most visible and transnationally active NGOs were chosen (*Climate Action Network International*, *Friends of the Earth International*, *Global Campaign for Climate Action* – also known as *TckTckTck* –, *Greenpeace International*, and *WWF International*). Taken together, this sample of cases, or actors, was thought to represent the major players and voices of both state and non-state nature in the global debate on climate change. This impression was confirmed in the course of fieldwork at the summit, where in addition to those in the sample, only few other actors could be detected as being particularly visible players in international climate PR. On the side of government delegations, they included the small island states, Bolivia, China, and – to a lesser extent – Canada, Japan, and Russia. Out of the NGOs, the only other organizations capable of attaining international visibility were Oxfam International and 350.org.

As required by my research design of structured, focused comparison, all actors had to be made subject to the same form of data collection. This meant that for each actor, a senior official involved in that delegation’s strategic communication decisions (like the respective communications director, chief spokesperson, or head of media relations) was to be recruited for a series of interviews. In that process, comparability was to be ensured by selecting interviewees who carried out comparable functions across delegations, even though carrying diverse job titles and placed at various positions in organizational structures. Recruiting of these professionals was generally successful (see 4.4.1.2), except for the delegation of the US.[11] In addition to these eleven actors in my sample, I also conducted an interview with a communication officer of the UNFCCC secretariat for the purpose of learning more about the organizational and logistical backgrounds of a UN climate summit.

11 For the case of the US, the first interview was conducted with a (rather uncooperative) communication staffer, who was only involved in public outreach, not in media relations. For the follow-up interview after the summit, I resorted to interviewing a journalist working for a specialized US news service and possessing profound experience in dealing with the US delegation’s PR at UN climate summits.

In general, I assumed for all interviewees that they were speaking on behalf of their respective organization, i.e. their government or NGO. I regarded them as representatives and mouthpieces of their organization and thereby treated 'actors' and 'interviewees' as essentially synonymous. I acknowledge the risk inherent in seeing complex organizations as unitary actors but – for the sake of reducing complexity – assumed that what PR professionals had to say was largely in line with official organizational positions. This does not mean, however, that I was insensitive towards the influence of personal backgrounds on communication decisions, as is shown in the empirical chapters.

OnlinePLUS: Due to their detailed technical nature, section 4.4 on data collection and section 4.5 on data analysis are not included here but can be downloaded from this book's companion website at www.springer.com.

4.6 Chapter summary

This chapter introduced the overall research design and empirical methods for this study. Political actors' communication strategies at the Cancún climate summit were investigated by means of a case study integrating elements of (1) *process-tracing*, the detailed reconstruction of concealed chains of influence on strategy choice, and (2) *structured, focused comparison*, the application of a standardized analytical framework – which emphasizes selected aspects of particular interest – to all actors under investigation. The sample includes the governments of six countries assumed to be highly visible in the climate change debate and five transnationally active NGOs. PR professionals from all eleven delegations were interviewed in the midst of summit business as well as afterwards by phone. In addition, observations of actors' communication structures and processes were conducted on site and an extensive corpus of actors' PR materials compiled. These supplementary data served purposes of contextualization and validation. Data analysis was carried out in form of computer-aided qualitative content analysis using the MAXQDA software package. The deductively developed coding frame was extended through inductive exploration of the data and the coding results investigated using different strategies of analysis.

5. Structuring the analysis and laying foundations

This and the following two chapters lay out the empirical findings of my case study of the 2010 UN climate summit in Cancún, Mexico. As stated in more detail in the previous chapter, the information presented here is first and foremost based on 25 interviews with communication officials working for governments, NGOs, or the UN, conducted at the event and post hoc by phone. Supplementary data stem from open, covert, nonparticipant observation on site (as well as impressions from a visit to the 2009 summit in Copenhagen) and a comprehensive corpus of PR material and conference documentation released at the summit. Original quotes from the interviews and PR material as well as relevant pictures are incorporated frequently throughout the chapters. Whenever information was obtained by different means than the ones described, the source is stated. All information presented here refers to the time of fieldwork.

5.1 'Dissecting' the summit: the structure of the analysis

While some sections of this chapter are descriptive in nature and simply portray particular structures and processes at the Cancún summit, others are more shaped by an explanatory or interpretative agenda. In general, the structure of the three empirical chapters is closely oriented towards the conceptual model introduced above (see 3.3.2). The underlying idea is to explore the model's different components and to assess their validity in a step-by-step manner – to 'dissect' the summit bit by bit. In the following, I outline the structure of my analysis; this can be seen as a 'manual' for better understanding why I arranged the findings in this particular order.

On the whole, the empirical part of this study is divided into three distinct components: Firstly, some background on the setting and infrastructure of the Cancún summit is introduced (section 5.2). While this is mostly about facts and figures, it provides necessary context for understanding the function of the summit as a public diplomacy resource. Subsequently, I investigate the communication structures and processes political actors had put in place at the event and their long-term, preconceived strategies (chapter 6). Also, the more stable factors influencing communication strategies, e.g., degree of professionalization, are examined and a typology of summit PR developed. Hence, this chapter is about the more *structural, stable, preplanned* aspects of carrying out strategic summit communication. The subsequent chapter is devoted to the more *procedural, dynamic, ad-hoc* side of the event, the short-term interactions between summit PR and summit proceedings (chapter 7). The specific unfolding of summit business might challenge political actors' communication structures and processes and cause adjustments to preplanned strategies; it might also provide actors with lessons to be taken into account in the future. Such responses to and interdependencies with the course of summit proceedings are explored in that chapter.

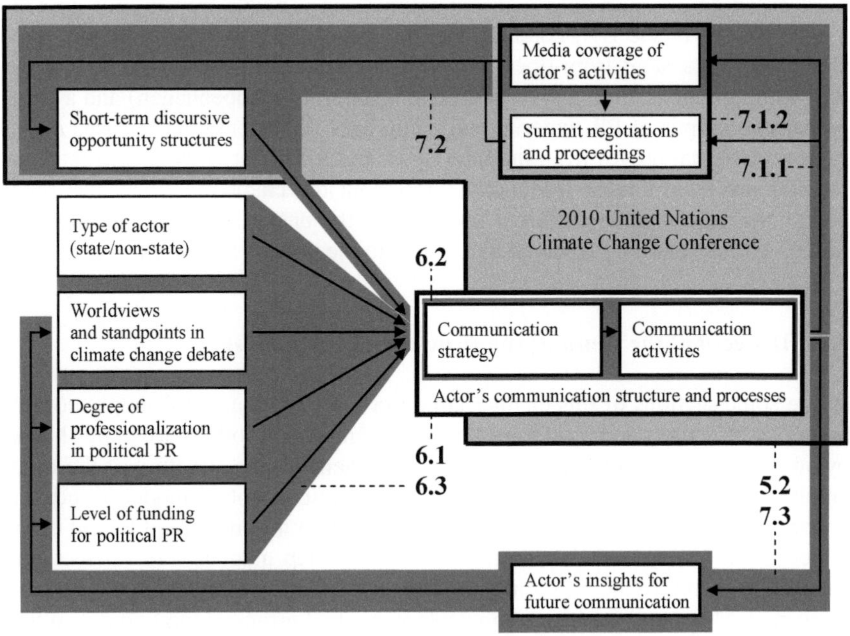

Figure 5.1: Links between chapter sections and components of conceptual model

This sequence of analysis is closely oriented towards my conceptual model (see figure 5.1): Section 5.2 introduces the parameters under which summit PR takes place, highlighted in blue in the figure. Chapter 6 brings summit PR itself into focus, i.e. this study's class of events, or phenomenon of interest: It explores political actors' communication structures and processes (section 6.1), then compares the communication strategies and activities that are carried out on the basis of these structures and processes (section 6.2), and also investigates the long-term factors shaping strategies (section 6.3). All these model components are highlighted in green. Finally, chapter 7 analyzes the two feedback cycles that are part of the model, the short-term and long-term mechanisms through which communication strategies might get adjusted: Firstly, strategies might influence the work of journalists at the summit (7.1.1), whose coverage, in turn, might influence summit proceedings (7.1.2). Such interconnections between summit coverage (influenced by summit PR) and summit proceedings might then open up short-term discursive opportunity structures for actors, which might again influence their strategies, as illustrated by an example (section 7.2). Besides these short-term influences, this chapter also looks at more long-term learning effects by actors themselves (section 7.3): Which lessons do they draw from their PR strategies and activities for prospective events? Did the specific unfolding of the summit cause them to reconsider their approaches in the future? Both feedback cycles – short-term and long-term – are highlighted in red in the figure.

The rationale behind this 'dissection' of the summit into individual components is to increase analytical quality. It allows for a structured perspective focusing on potential causal mechanisms operating at the summit and is essentially a prerequisite for conducting the approach of 'process-tracing' described in chapter 4. While such a step-by-step procedure might entail some minor overlaps at times, it provides for a more valuable look at the summit's public communication procedures. But before these procedures are brought to the fore, a look at the general setting and infrastructure of the summit seems sensible.

5.2 Setting and infrastructure of the 2010 UN climate summit

As the 16th summit within the UFCCC process, the 2010 UN Climate Change Conference (COP-16) took place from November 29 to December 10 of that year. This section introduces some background on who organized and participated in the event (5.2.1), at which location and particular venues it took place (5.2.2), and which infrastructure it provided for the work of PR professionals and journalists (5.2.3).

5.2.1 *Organization and participation*

The organization of a UN climate summit is the product of close collaboration between two actors: the UNFCCC secretariat, based in Bonn, Germany, and a respective host government, i.e. that of Mexico for COP-16. The UNFCCC requires host governments to meet a comprehensive set of requirements when hosting a summit, ranging from the provision of certain facilities and the offsetting of carbon emissions to the display of UN flags on the premises and assurance of police protection (Government of the United Mexican States and UNFCCC Secretariat, 2010). All expenses created by not holding the event at the UNFCCC's seat in Bonn are to be covered by the host government. Governments wishing to host have to express their interest in advance and build support for their candidacy within their regional group (such as Africa or Latin America and the Caribbean), which subsequently decides about its candidate. By convention, the regional groups take turns in hosting. The host government of a given COP is formally elected at the preceding summit.

When hosting a COP, the host government commonly assumes the presidency of that summit and designates one of its ministers – usually the minister of the environment or the minister of foreign affairs – to be the COP President, who chairs the most important sessions and may assume a central role in shaping the negotiations and facilitating agreement. The President is supported by the Executive Secretary of the UNFCCC, who heads the secretariat and watches over the general political process. At Cancún, this dual leadership comprised COP President and Mexican Minister of Foreign Affairs Patricia Espinosa and UNFCCC Executive Secretary Christiana Figueres. One interviewee saw the good chemistry and emotional fit between both women as one of the factors behind what he regarded as the eventual success of the talks.

5,192 participants delegated by governments – or 'parties', in UN language – took part in the summit. They were joined by 5,386 participants from observer organizations and 1,270 journalists (UNFCCC, n.d.). This adds up to 11,848 participants in total (the preceding summit in Copenhagen saw more than twice as many). 'Observer organizations' refers to all actors monitoring but not formally taking part in the negotiations, such as NGOs, research organizations, or religious groups. The UNFCCC process has traditionally been very open to non-governmental actors (see 3.2.3), a UN communicator pointed out:

> "Compared to other negotiations, like WTO, for example, we have an open-door policy. […] NGOs are part of the game. They can raise their voice; they can participate to a certain extent. Of course, decisions are only made by governments; however, they have a big stake in the whole negotiation process" (3321, p. 9, 11).

For observers not familiar with the process, the high visibility and established roles of NGOs at UN climate summits may indeed seem remarkable. Around a core of highly technical negotiations, a colorful periphery of civil society representation has emerged, utilizing the platform of climate summits for monitoring, lobbying, and campaigning. This is formally appreciated by the UNFCCC, although certain rules have to be adhered to.

For example, in Cancún, if an NGO planned to carry out a public activity, such as a larger photo op or an on-site demonstration, the UNFCCC's NGO team had to be notified two days in advance. It was then assigned a particular time and location for its activity. According to the UNFCCC, this rule was to guarantee order on the premises (local police had apparently specifically asked for such a regulation) and to allow for a fair distribution of protest opportunities among NGOs. Their PR professionals, however, largely criticized this rule for obstructing quick-paced forms of campaigning, as one NGO communicator described:

> "If a country makes an outrageous announcement and you want to expose that and highlight that and you can only do so two days later, the story is gone! I think it really was a problem and restricted the possibilities of NGOs to express their concerns by those means. We could still write press releases, but visual actions got more and more difficult" (2176, p. 8, 22).

The introduction of this requirement for advance notices in Cancún might also be seen as an attempt by the UNFCCC to find ways of channeling NGO activity more explicitly in response to some chaotic situations at the previous summit in Copenhagen. Some NGO communicators described the requirement as a new lever on the side of the UNFCCC for potentially limiting NGOs' disturbance of the talks (similarly to the separation of venues described below). However, with the exception of one smaller incident involving minor disturbances and interventions by UN Security, COP-16 remained a quiet and peaceful summit.

5.2.2 *Location and venues*

While up until early summer 2010, some coverage had named Mexico City to be the likely site of COP-16, the government of Mexico selected the coastal city of Cancún on the Caribbean shoreline of Yucatán peninsula as location. Extensive tourist infrastructure, numerous direct flight connections from around the world, and experience in hosting large events made the city an obvious choice. The summit itself took place at two separate venues outside the southern suburbs, right off the federal highway connecting the city to the airport and the neighboring city of Playa del Carmen.

The actual negotiations were situated at the hotel compound Moon Palace, located directly at the coast of the Caribbean Sea. The hotel provides more than 2,000 rooms as well as several event halls of different dimensions (Moon Palace, n.d.). For the event, the entire 500,000-square-meter complex was turned into a convention center, complete with two plenary halls, numerous meeting and briefing rooms, a media center, working zones, as well as restaurants and snack bars. Many of the suites were converted into offices for government delegations or UN representatives, while some also remained available for individual accommodation. The whole of Moon Palace was surrounded by heavily armed security perimeters.

The second part of the summit was located at Cancún's newly erected exhibition halls named *Cancunmesse*. The halls accommodated what has become an integral part of climate summits: a colorful array of governments' and NGOs' information booths as well as larger country pavilions. Furthermore, the exposition grounds provided rooms that governments or NGOs could reserve for briefings, talks, panel discussions – the so-called 'side-events', in UN language –, or for internal meetings. Registration and information desks, a large food court, and the central arrival and departure area were also located here. With every arrival at the summit, participants – except for VIPs and their entourages – had to undergo airport-like security screenings at *Cancunmesse*, even if Moon Palace was their desired destination. Also, leaving the summit was only possible through the expositions grounds. The two venues were connected through frequent shuttle buses that took about twenty minutes for the transfer.

Due to the remote location of the summit venues and the lack of efficient public transport in Cancún, the organizers had set up an elaborate network of bus routes stretching across the metropolitan area for the duration of the summit. It connected all official hotels with the summit gateway at *Cancunmesse* and also integrated the Climate Change Village. This was an exhibition park set up by the Mexican government in the suburbs of Cancún, where it presented its patriotism-laden *México en tus sentidos* exhibition, gave domestic enterprises and NGOs the opportunity of showcasing their climate action, and held concerts and other events – all of this mainly geared towards the local population. Travelling on the exclusive bus network sometimes required patience as well as changing buses at designated transfer points. This was frequently pointed out as inconvenient by summit participants.

Especially the delay caused by commuting between *Cancunmesse* and Moon Palace was met with heavy criticism. Some PR professionals even posited that the frequent bus rides slowed down their on-site communications work and saw the splitting up of the summit as an obstacle to seamless coordination with colleagues. The separation of venues certainly stood in clear contrast to the setting of the Copenhagen summit, where all event components were brought together inside the Bella Center. However, the close proximity of NGOs and negotiators there also led to some chaotic protest situations that seemed to catch organizers by surprise and were criticized by governments. As with the introduction of new NGO rules mentioned above, the particular way of spatially organizing the summit in Cancún (and essentially breaking it up into a governmental and a civil society 'sub-summit') can be seen as an attempt by the UNFCCC and the host government to calm down the negotiation setting and, in the case of chaos, be able to exclude civil society actors from accessing the negotiations. This was even met with some understanding on the side of NGOs:

> "They have to learn from Copenhagen. I guess that it was probably a convenient way to keep things together and yet being able to separate them. [...] It is inconvenient for us – especially when you have a two hours' drive to come here –, but it isn't stupid. I understand why they do it" (6282, p. 4, 7).

Other PR professionals speculated that the relationship between the two core components of UN climate summits – the official negotiations on the one hand and civil society representation on the other – will always be subject to some adjustments and experimenting.

5.2.3 *Communication infrastructure*

At any COP, the UNFCCC, together with the host government, sets up a comprehensive infrastructure for media representatives to report on summit developments as well as for political actors to express their positions vis-à-vis the media. The hub of media activity is commonly found at a dedicated media center that can only be accessed by accredited press. Governmental and nongovernmental participants are specifically banned from the facility (with minor exceptions) in order to provide journalists with a refuge for uninterrupted work, as a UNFCCC public information officer explained:

> "This is the last resort for the media to be in a neutral area [...]. Because when you come from the media – you have seen this in Cancún or Copenhagen –, there are so many people who approach you, because they want to sell their story, whatever they have to sell. You have to allow the media to digest all of this and to write, to work, to produce in their own area. That's why we don't allow access to NGOs, we don't allow access also not to governments" (3321, p. 13, 4).

In general, the provision of communication infrastructure serves the UNFCCC's explicit motive to facilitate media coverage of the climate summits and make transparent the international political efforts in combating climate change. While the mandate of the UNFCCC does not allow for large-scale strategic communication efforts, the secretariat does operate various smaller initiatives in the field. Particular attention is given to journalists from developing countries, who often lack the resources and knowledge to adequately report on COPs. In order to enable them to attend – and also to sensitize them for their countries' particular vulnerability to climate change –, the UNFCCC, in cooperation with selected governments and foundations, runs a program covering the travel expenses for these journalists and training them in the fundamentals of climate change. This is part of a deliberate effort to make the communication infrastructure at COPs available to a set of actors as diverse as possible.

At Cancún, the media center – as the backbone of that communication infrastructure – was located in a distinct building within the Moon Palace complex. It was separated from the epicenter of summit activity by a five-minute shuttle bus ride. This isolated location – at the previous COP, the media center was set up right in the heart of Bella Center – was met with strong criticism by journalists and recognized as logistical lapse by the host government itself:

> "We had some problems dealing with the distances […] between the press area and the place where things were taking place. So the press had to walk or take a bus or else, travel a long way to get to where the information was. […] It was a certain issue that was of importance, and we were quite worried about it, because the thing wasn't flowing because of the huge distance, logistically speaking. […] There were complaints [by journalists]. We had to establish some other extra spaces for them to work" (0240, p. 1, 20).

As a consequence, parts of the press corps got used to working out of general working zones located closer to the negotiation rooms, where they also found themselves to be in proximity to NGOs' PR professionals (see 7.1.1.3).

The media center comprised the international broadcasting center (IBC), an area reserved for writing press (equipped with Internet access and printers), and stand-up positions, where TV correspondents could record their statements. Especially the setting up of the IBC requires long-term planning that, in the case of Cancún, began about ten months prior to the summit with what is called a first 'technical mission' in UN language: UNFCCC media staff meet with their counterparts in the host government, clarify the latter's motivation to host the event and how this has to be considered in the setting up of communication infrastructure, inspect the proposed venue, and provide advice in selecting a host broadcaster. Additional technical missions might follow throughout the year. UNFCCC media staff arrive at the summit venue about two weeks prior to its beginning to monitor set-up. While the staff usually comprises around eight members, for the occasion of COPs, additional media staffers from other UN agencies are requested in order to fulfill the extra demand for personnel caused by accreditation and press conference supervision.

Host broadcasters play a significant role in the construction of a summit's media image, as they are the exclusive provider of footage of the official proceedings in the plenary halls (with some minor exceptions). While TV crews may record their own material anywhere on the premises and in press conference rooms, for the actual negotiations, they depend on the images – and image quality – made available by the host broadcaster. While the host broadcaster is usually one of the major TV stations in the host country (such as Danish station TV2 for COP-15), in the case of Cancún, a production company was tasked with assuming that role. In addition, Eurovision Americas, the Washington-based US subsidiary of the European Broadcasting Union (EBU), was put in charge of worldwide image distribution.

Host broadcasters also run the technical infrastructure at the IBC. At Cancún, the IBC was set up the 'North American way', providing TV stations with open and sparsely furnished cubicles instead of proper offices that could be locked. The host broadcaster would service TV crews' cubicles, supply them with the different signals from around the summit venue, and handle the distribution of video reports via satellite uplink or – as a more recent development – broadband. The latter service may involve fees that are to be covered by TV stations themselves (as for any other additional service used, such as the provision of studio facilities, the renting out of ENG crews, or the installation of additional phone or fax lines).

In addition to the media center/IBC, the communication infrastructure also included dedicated press conference rooms, which governmental and non-governmental actors could reserve for holding media briefings. These rooms were professionally equipped with a stage featuring a pedestal, a panel desk, and a backdrop showing summit and UNFCCC logos. Every room was staffed with an engineer handling sound, stage lighting, and the filming of proceedings in that room. The video feeds from press conference rooms could be accessed at the media center and TV sets installed all over the premises. At Cancún, 253 media briefings were held in two dedicated press conference rooms, one of which accommodated governments (and some selected NGOs), while the other one was reserved for NGOs and IGOs. Only accredited media representatives were allowed in the press conference rooms, although this was not always checked for the NGO room. Generally, governments have priority over NGOs when requesting a room; should clashes occur among governments, UNFCCC media staff attempt to exercise diplomatic sensitivity.

Finally, the communication infrastructure at Cancún also comprised a new type of facility specifically set up for bloggers and NGO communicators. This so-called 'Bloggers' Loft' featured a dedicated working zone equipped with special bandwidth facilitating online work, especially video upload. Through the introduction of this facility, the UNFCCC responded to the emergence of a new set of 'grey-zone actors', who report from the summit but could not be accredited as media, as they lack the necessary credentials or are tied to political actors. Especially NGOs have built up significant media production capacity over recent years, sometimes arriving at the summits with entire camera crews. In an effort to provide these new hybrid actors engaging in activist 'narrowcasting' with an appropriate infrastructure outside the media center, the Bloggers' Loft was created. This, again, might point to the UNFCCC's rather NGO-friendly attitude, which in the past could also be seen in some informal acts of support by UNFCCC media staff to niche NGOs overwhelmed by summit business. In such cases, the UNFCCC might provide some guidance on drafting press releases or utilize its contacts to journalists. Such support, which the UNFCCC itself calls "capacity-building" (3321, p. 12, 9), is always granted on an exceptional basis and limited to smaller actors with little experience in doing public diplomacy.

5.3 Chapter summary

This chapter laid the foundation for the following two chapters. It firstly spelled out how the structure of my case study is closely oriented towards the conceptual model developed in the theoretical part. Empirical testing of this model proceeds in a step-by-step manner, 'dissecting' public communication processes at the summit bit by bit. This provides for analytical value and accommodates the methodologies of structured, focused comparison and process-tracing. Secondly, the chapter introduced the organizational and infrastructural parameters of COP-16. It presented the summit as an enterprise collaboratively organized by the UNFCCC and a respective host government and highlighted organizers' efforts at facilitating the work of journalists and PR professionals. The establishment of well-equipped media centers, media briefing rooms, and online PR facilities can be regarded as indications for the UNFCCC's agenda to promote coverage on international political efforts against climate change and to nurture civil society involvement. It is also these deliberate measures by the UNFCCC that turn climate summits into facilitators of public discourse around the world and strategic resources for public diplomacy. This is an intended function of such events, and subsequent chapters investigate how political actors respond to this intention.

6. Political actors' communication structures, processes, and preplanned strategies

This chapter outlines how political actors, both of state and non-state nature, utilize the resource of the 2010 climate summit for their communication efforts: Which communication structures and processes do they build onto this platform (section 6.1)? Through what kind of communication strategies and activities do they exploit its worldwide visibility (section 6.2)? The findings of these two sections are summarized in tables 6.2 and 6.3. Subsequently, the third section (6.3) explores to what extent strategies are shaped by actors' more long-term properties, such as their worldview or degree of professionalization. Finally, this chapter introduces a typology of the different approaches of summit PR found in Cancún (section 6.4). Overall, this chapter focuses on the structural and stable aspects of summit PR.

OnlinePLUS: Due to its detailed technical nature, section 6.1 on political actors' communication structures and processes is not included here but can be downloaded from this book's companion website at www.springer.com.

6.2 Political actors' communication strategies and activities

While section 6.1 was actor-centered and laid out my findings on communication structures and processes at the Cancún summit in a rather descriptive manner, this section is more analytical in nature. Focusing on different dimensions of strategic communication, i.e. messaging (6.2.1 and 6.2.2), media relations (6.2.4), and other PR activities (6.2.5), it compares and integrates findings across actors. In statistical language, we are now switching over to a perspective focusing on the columns, i.e. variables, in a dataset (as opposed to the rows, i.e. cases).

6.2.1 Governments' messaging strategies

When governments and NGOs engaged in strategic planning of summit PR, this strategizing often concerned the overall message or theme they wanted to bring across at the event. On average, NGOs appeared to invest more effort in such preparations, sometimes developing comprehensive internal memos laying out core messages and corresponding talking points. These guidelines were to be observed in all communication activities, be it the briefing of journalists or the staging of photo ops. On the side of governments, less sophisticated planning of messaging strategies could be found; some countries appeared to have come to Cancún without specific plans for which themes they want to emphasize in summit PR.

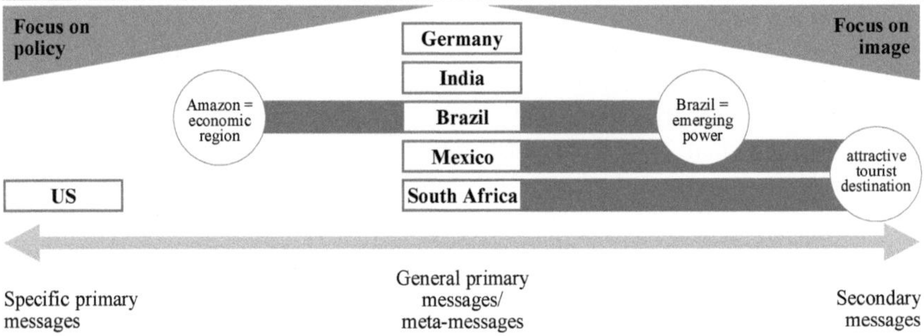

Figure 6.1: Spectrum of governments' messaging strategies

Generally, governments' Cancún messages can be classified according to the degree to which they referred to the specific context of COP-16 and to the current state of climate talks. Hence, at the one end of the spectrum (see figure 6.1), we find (1) *specific primary* messages dealing with particular positions in international climate negotiations. The middle of the spectrum accommodates (2) *general primary* messages, which relate to the climate talks but do not convey a detailed negotiation position. A special form of such messaging is constituted by *meta messages*, which concern the way in which the talks are conducted, rather than their political substance. Finally, at the other end of the spectrum, (3) *secondary* messages are found. Such messaging does not refer to international climate policy but exploits the summit as general communication resource. Here, messages might emphasize a country's qualities as holiday destination or candidate for foreign investment; the basic idea is to profit from global attention to the climate summit for non-climate-related purposes (for more on the difference between primary and secondary messages, see 3.3.1). Hence, as we move down the spectrum, governments' messaging focus changes from policy to image.

While all government delegations more or less specifically commented on the progress of the negotiations in their media briefings, their overall messages emerging across the full range of communication activities and acknowledged as such by their PR professionals clearly differed. In the case of the *United States*, however, it was not possible to identify such umbrella message as backbone of summit communication, as most substantial content was released by the two senior negotiators in media briefings and naturally related to specifics of the talks, such as the US delegations' more resolute position vis-à-vis China. In the absence of more overarching messaging, this is what we have to count as the US delegation's message.

Most other governmental delegations supplemented their specific commenting of the negotiations with top-level messages that can be located around the middle of the spectrum. While they referred to climate change, they did so in a rather abstract way by pointing to the general 'responsibility' the country is taking or the 'homework' it is doing in that regard. For example, the overall message of *Brazil*'s PR at Cancún was that the country is a leading player in sustainable development and, by giving itself ambitious policies, is contributing to the global effort to combat climate change. In a specific twist, however, Brazil's messaging also referred to the Amazon region, which the country regards as unique wildlife reserve but also as a place where people live and work. The Amazon should hence not be fenced off and excluded from economic development, as the head of Brazil's public diplomacy explained:

> "We still have a problem with the Amazon – people in general don't have a comprehensive view of the Amazon – so that's why we've been trying to show that Amazon is not just a closed forest but also that there are people living there. And they are trying to solve their economic problems, trying to get out of poverty, and we have many projects of sustainable development in course in the Amazon" (6775, p. 3, 31).

This message can certainly be described as a more complex one, as it touches upon the fine line between utilizing the Amazon as place for settlement and source of resources on one hand and exploiting it on the other hand. Accordingly, the exhibition at Brazil's ornate country pavilion, *Espaço Brasil* (see 6.2.5), showcased products and cultural traditions coming from the Amazon region but also highlighted projects by businesses and civil society to preserve nature as well as political efforts to prevent its destruction. A visitor was to realize that, yes, Brazil knows the Amazon is in danger and does something about it but, no, will not "close down the Amazon and throw the key in the Atlantic Ocean" (6741, p. 9, 26). Setting up the pavilion as dedicated platform for explaining this delicate argument seemed sensible. This self-confident approach, together with the general professionalism and opulence of Brazil's summit PR, seemed to subtly hint towards a secondary message: that Brazil is an emerging power with high ambitions on the world stage.

Just as Brazil emphasized that "we are doing our homework" (6741, p. 2, 22), other countries also conveyed general primary messages referring to their assumption of responsibility for combating climate change. In the case of *India*, the environment minister's aide described India's overall communication strategy as comprising two components. One was targeting the international arena and aimed at making clear that the country was assuming more responsibility for climate change than it had done in the past:

> "We are a responsible nation. We recognize that, although our per-capita emissions might be low, we are increasingly a large developing country with growing emissions, and therefore we cannot just sit at the boundary and say 'we will not do anything'. For the sake of the world, for the sake of the global community, we will act!" (6468, p. 1, 18)

> "We want to tell people [...] how we are being much more proactive than we used to be on climate change. For that, targeting the international media was very important" (6468, p. 5, 6).

The second component was targeting the Indian population and aimed at ensuring that "that people [...] know what's going on, that there is no misinformation, that people understand the context in which the minister is making his decisions and statements" (6468, p. 5, 8). Both strategy components were linked, as India's seemingly selfless and self-committing appearance at the international level might have been viewed critically by its own population, which could have suspected its government of selling out national interests.

As unspecific as India's international message might appear, it does convey a sense of confidence and eagerness to be respected on the world stage, similarly to Brazil's approach. Outside the BRICS world, *Germany* also chose to adopt an abstract responsibility message but did not appear to connect it with particular communicative ambitions. Instead, the idea was to generally emphasize the progressive role Germany plays in environmental protection at home and abroad and to stimulate conversation with interested members of the environmental community about these cases of best practice. While German communicators admitted to 'Germany takes responsibility' as being a rather vague motto, the manner in which this message was communicated also appeared somewhat uninspired, as if it was not clear why it was chosen in the first place. In exhibiting such reluctance to utilize the summit as resource for strategic communication, Germany came close to the US delegation, whose summit PR appeared to lack an overarching messaging approach altogether.

As hosts of the Cancún summit and its 2011 successor in Durban, the governments of Mexico and South Africa aimed at conveying a different form of message, which is located closer to the 'image' pole on the spectrum introduced above. Their hosting, or prospective hosting, of the UN climate summit was tied to an explicit public diplomacy agenda of improving the national image and recommending themselves as serious political and economic partners. In the case of *Mexico*, the logistical architect and political facilitator of COP-16, this was principally done through emphasizing the process of the negotiations, the way in which they are led by the Mexican presidency. While the UNFCCC commonly tries to implement an approach of transparency and openness at COPs (with varying degrees of success), the Mexicans explicitly picked up on this notion and expanded it to become the grand theme of their COP and the core message of their summit communication. In clear contrast to the heavily criticized decision-making in unofficial negotiation circles and behind closed doors at the preceding COP in Copenhagen (Hunter, 2010, pp. 12-13), the Mexican government aimed at making the entire talks as open and engaging as possible. This theme dominated nearly everything the delegation publicized in Cancún – even up to a point where it became unclear where the Mexicans stood in terms of solid negotiation positions, as another delegation's senior communication official pointed out. Transparency had become the exclusive message. This was part of a conscious strategy, as the Mexican Director of International Information explained:

> "Since we knew that we were going to be hosting this and [Foreign Minister Espinosa] was going to be the president of it, she always talked about and showed and gave the example of the transparency, in which this conference should be managed" (0233, p. 2, 31).
>
> "She, along with the Executive Secretary, has always made the compromise of letting everyone know what's going on – through press briefings, through interviews, through every way [possible]. So whenever there is someone saying that there is something behind doors and something under the table, they go and say 'no, this is transparency', and we are actually using a real transparency issue as the best way to communicate this" (0233, p. 3, 12).

The Director of International Information, who stressed the theme of transparency repeatedly in our interviews, explained the emergence of that message also with the character and background of COP president and Mexican foreign minister Patricia Espinosa:

> "This is the first time that a minister who is not from the environment [ministry], who is foreign minister, is presiding over the COP. So that changes the whole situation, because it's not a matter of the substance only, but of the negotiations and the way you care [about] and handle things between governments and within the negotiations: The way in that a minister of environment deals with those things is quite different than what a minister of foreign affairs does – also because our minster of foreign affairs is a diplomat of career, and she is a member of the Foreign Service, and she has been doing that her whole life" (0233, p. 3, 4).

Large parts of media coverage on the conclusion of the Cancún Agreements at the end of the summit seemed to attribute the reaching of a consensus (not counting Bolivia's objections) to Espinosa's transparent chairing of the negotiations. In some media coverage, she was even heralded as a "goddess" for her negotiation skills (Schwägerl and Traufetter, 2010). The overarching theme of transparency had apparently found its way into media coverage.

Behind the promotion of this meta message seemed to loom a more fundamental interest on the side of the Mexican government to present to the world a different side of the country. Hosting the COP – and doing so in an inclusive and widely respected way – would instigate favorable positive coverage about Mexico, government communicators reasoned. A UN official speculated that the government wanted to distract attention from the drug war in Northern Mexico and put out a counter-image of Mexico as attractive tourist destination:

> "What you could read between the lines was also [the plan] to sell Cancún as a conference destination, as a holiday destination, because you produce two weeks of pictures, images from a conference. […] When you asked them, for example [regarding] the stand-up positions, 'where do you want to you have your stand-up positions', they said 'oh yeah, there should be a palm tree, a beach, the sea'. So they want to transport also then an image of a city, country! And when we had the first meeting with the officials in Mexico, they had this… they still have the problem in the North of Mexico with the drugs and the trafficking […]. And I got the feeling that they wanted to counterbalance a little bit the reputation of the country. So it is not only the war on drugs – Mexico is also a beautiful holiday destination" (3321, p. 3, 18).

This impression was confirmed by the Director of International Information, who concluded after the summit that the image of Mexico had been altered slightly:

> "We had a boost of good news about the country, about what is Mexico beyond these issues of violence and narcotics" (0240, p. 1, 3).

> "I do think that beyond all the rest of the information that the world gets about Mexico, certainly COP-16 gave another kind of information […]. It's not only about drugs and drug traffic and violence and everything, but also a place where consensus can take place" (0240, p. 3, 12).

Overall, Mexican messaging centered on themes of transparency and Patricia Espinosa doing a respectable job as COP president and, through this, being a symbol for the positive sides of Mexico beyond the drug war. Hoping for similar image effects, *South Africa* commonly emphasized in its summit communication the country's ability to be a good host for COP-17 and an equally attractive holiday destination. It was argued that COP-17 would bring to Durban a particular spirit of excitement, which could as well be enjoyed for some additional holiday days after the summit. But besides this blunt secondary messaging, South Africa's core message centered on the climate threat's gravity, especially for developing countries:

> "Our priorities and our resources are channeled towards ultimately eliminating poverty. And with climate change and the threat that it poses, especially with regards to agriculture […], it has a potential to threaten […] the ability of the African continent to feed itself. […] So our message is that we need to look at climate change as more than just an environment issue. It's a sustainable development issue!" (1807, p. 2, 12).

At the same time, in an extension to this general primary message, the South African communicator pointed out how the next COP in Durban could provide for conditions conducive to combating this serious threat – politically and logistically:

> "Durban promises to be a platform that nations of the world should look forward to in terms of the climate change talks" (1807, p. 7, 22).

> "I think with that good image of the World Cup having come up in 2010, it already showed that we are able to host events of international stature. And I think that was a plus for us" (1894, p. 4, 34).

> "We've been quite clear: we are ready to host the COP" (1894, p. 4, 23).

In an interesting twist, South Africa at the same time attempted to manage expectations and avoid the mistakes of COP-15 in Copenhagen, which fell apart under the high hopes and strong symbolic meaning attached to the summit:

> "We really wouldn't want to create unnecessary expectations. […] I know that there is a lot of expectations, and justifiably so, but we […] need to be

> very practical, we need to be very honest on what's possible and what's not possible, what's doable in the context of the issues at stake" (1894, p. 2, 11).

Hence, South African messaging at Cancún featured a general primary message relating to the severity of climate change – this was mostly oriented towards domestic audiences – and a cautious meta message as part of deliberate expectation management. Also, the country was presented as attractive tourist destination.

6.2.2 NGOs' messaging strategies

The motivation to shape public expectations surrounding a climate summit, which played some role in South Africa's communication strategy, can be considered *the* driving force behind the messaging approaches that NGOs carried out in Cancún. COP-16 saw the roll-out of several new communication strategies that NGO alliances and individual NGO had never carried out before and specifically responded to the grand disappointment of COP-15, which, despite high hopes, had failed to reach an agreement. Coordinated by CAN International and GCCA, large parts of the NGO community decided to approach Cancún with a down-to-earth perspective and adopted messaging strategies that were described with phrases like 'staying modest', 'setting expectations right', or 'being realistic' (see figure 6.2 on page 129).

In response to the failure of Copenhagen, *GCCA* had started an internal discussion about how to continue its campaigning effort. In a process involving many member NGOs, it was decided to abandon the former approach of overloading the climate summit with expectations and of presenting an international treaty as the only means of saving the climate. Instead, in a reworked messaging strategy, expectations were deliberately lowered and attention shifted away from the negotiations:

> "After Copenhagen, I think we all saw how dangerous it is if you rely only on the UNFCCC and you don't have an alternative narrative that kicks in or that runs parallel. We all fell into a big black hole, there was lot of depression, the movement was in tatters, and the skeptics and all those who have a different narrative, an anti-climate action narrative, had a field day, because they felt that they could fill this vacuum with their story. So we are trying to tell the story that is bigger than the UNFCCC, in which the UNFCCC is not an end in itself but more a means to an end. […] There is more happening, don't just rely on that treaty!" (2135, p. 4, 21)

This approach of "telling a different story of this climate debate, reframing climate change, and reaching new and bigger audiences" (2135, p. 3, 14) was built mainly on making clear to the public that climate action was not such much about one grand political project, but more the sum of many small personal projects undertaken by individuals around the globe:

> "We feel that climate action as it happens is already much broader than the UNFCCC. A lot is going on out there that happens no matter what happens here! And we find it important to talk about this and to amplify and aggregate these stories. Because there is a wrong perception out there – media love to talk about it, people then read this in the media and think this is the truth – that if this process doesn't deliver, we are all doomed and nothing else is happening. That's simply not true!" (2135, p. 3, 33)

According to GCCA, this new framing, which debuted at Cancún, should be seen as work in progress and will certainly require more fine-tuning and discussion with member NGOs. The strategy of pointing out the significance of individuals' small steps in dealing with climate change was condensed in the catchphrase 'race to the future', which served as the backbone of GCCA's communication activities. The strategy was clarified and broken down by means of an internal five-page document that GCCA had developed prior to Cancún in cooperation with communicators in member NGOs.

While GCCA aimed at drawing attention to those forms of climate action taking place outside the UNFCCC, *CAN International*, with its focus on directly influencing negotiations, attempted to take some pressure off the talks and adopted messages that were appreciative of smaller political steps:

> "You need to think realistically about what is achievable at each negotiation. So our messages were around those lines, where we were talking about building blocks, we were talking about victories on a more realistic – or less Big Bang – level that could set us on a path towards the full ambitious agreement that we are looking forward to in the future" (3172, p. 3, 13).

On the one hand, this pragmatic approach seemed to contradict established modes of voicing ambitious demands:

> "Of course it's a little depressing that that's what we are calling for, because you always want more" (3172, p. 4, 1).

On the other hand, it was viewed as the only way of securing at least some results in Cancún and not reliving the experience of Copenhagen, where the summit collapsed under the tremendous weight of expectations projected upon it:

> "So our communications were around setting expectations, a level that we thought was both ambitious but realistic, and communicating that to both governments as well as the public, so that there were the right level of expectations on this meeting, so that we could ensure that we didn't get caught in a situation where we were calling for something that was totally not possible and therefore doom the talks for failure before they even started" (3172, p. 3, 19).

In accordance with the reworked strategies carried out by GCCA and CAN International, *Greenpeace International* also tested a novel messaging approach in Cancún. In place of traditional strategies of denouncing governmental efforts and always calling for more, the NGO went for optimistic, empowering frames, much in line with GCCA's 'race to the future' and CAN International's 'building blocks'. This approach of 'going positive' was proposed to Greenpeace International by a London-based communication consultancy. Internally, it was seen as a significant break with older campaigning traditions:

> "We've campaigned quite negatively, used quite negative messaging, for a long time, and we think it's time to really talk about hope – some useful statistics, nice little stats in there, like 'over the course of this meeting, China will have built 300 new wind turbines'. So it's about opportunity. We don't want to dis them [i.e. the politicians]. Copenhagen was so dim, was so dark and bleak, and everyone lost hope – and cold and just tragic. And everyone lost hope. It's about keeping our supporters and keeping people thinking that it can happen. It's possibility! Our future is possible!" (5356, p. 4, 9)

'Choice' and 'opportunity' were the keywords of Greenpeace International's messaging strategy. As already hinted at in the quote above, it was directed at two audiences: On the one hand, governments were to be given the signal that progress in negotiations simply depends on their willingness to act and that, in this, even small steps matter:

> "To continue to bash the governments, just bash them, doesn't actually move them forward. They'll just go 'well, we tried, we got something, and then, oh, we just get bashed for it'. And I think giving the governments – even though they don't bloody deserve it – a bit of breathing space to actually get some work done, which they did, and to praise them for that small amount of work that they got done, was quite important. We didn't praise the whole thing. I was very pleased with our line on it, that 'we saved the process but we haven't yet saved the climate', which didn't slam quite so badly but talked about some of the stuff that we did get" (5337, p. 3, 22).

On the other hand, the new messaging strategy rolled out in Cancún was also to give wider audiences around the world assurance that some cautious optimism is justified and that they should not yet write off the idea of saving the climate:

> "If you completely slam the process again, you completely slam everything, [...] people will give up on any action around climate change! It's also that side of things; you just don't want to completely turn off all the members of the public, who are just going to be so despondent" (5337, p. 3, 32).

While *WWF International* kept a rather low profile at the Cancún summit and, instead of vividly promoting a clear message, gave preference to briefing journalists on the backgrounds of the talks, its general rhetoric was still compatible with the novel messaging approaches of the alliances. The Head of Communications and Campaigns in WWF's Climate and Energy Initiative saw the summit as the moment "where they really need to get going" (6280, p. 4, 1), while it would be too early to except a full-fledged treaty to be passed. He advocated messaging that also accepted small political steps as signs of momentum in the talks:

> "Up to Copenhagen, the idea was to bring in one big revolutionary new piece of legislation, the 'fab' [= fair, ambitious, and binding] treaty and use that to trigger the economic change that we need. And now it's back to square one. And we have to do all the squares individually and work on the economic change in each of the different areas. [...] And the treaty will come later, when the world is ready." (6297, p. 6, 23).

Similar levels of pragmatism and optimism were also present in NGOs' concluding messages at COP-16, in which they offered their evaluations of summit outcomes. As mouthpiece of most NGOs, GCCA tried to promote a balanced assessment of the Cancún Agreements summit communiqué, arguing that while substantial steps towards rescuing the climate were missing in the text, it had at least proven the capability of governments to act and build a base for future decisions:

> "[We did not go] for a simple 'success or failure' PR approach but tried to explain why this was an important step forward in terms of the political process being saved and a basis for further steps in the right direction being laid et cetera, but at the same time acknowledging that despite those positive developments that we did not necessarily expect to see, we did not come much closer to substantial agreements and to reducing emissions and to saving the climate" (2176, p. 2, 31).

> "While you have political demands that are higher or stronger or more ambitious than what came out of Cancún, you also know that you need a cer-

> tain momentum to get anywhere and that it would be much more difficult to get what we want if Cancún had ended in collapse. The fact that it did not end in collapse led these groups [i.e. mainstream NGOs within GCCA and the GCCA secretariat itself] to take this balanced PR approach" (2176, p. 4, 11).

By taking this stance, GCCA saw itself in a middle position between advocates of a rather pessimistic assessment of the Cancún Agreements, such as Friends of the Earth (see 6.2.3), and some governments whose reactions were close to enthusiasm. It was also compatible with one particular communication objective that GCCA had given itself for the Cancún summit: to make clear that the UNFCCC process is not flawed by nature but that it is constituted by governments possessing considerable leverage in making it work. Bringing that point across was a prime motive in GCCA's final Cancún assessment – as was the highlighting of the impact of individual, non-UNFCCC climate action within its general messaging strategy.

6.2.3 A closer look: Dissent over strategy among NGOs

One NGO in my sample did not adhere to the messaging strategies developed under the auspices of GCCA and CAN International: Friends of the Earth International. Behind this fact seemed to loom a deeper separation between two camps in the climate movement. On one side were NGOs that described themselves as 'mainstream' or 'pragmatic' and viewed climate change as a challenge that had to be dealt with by all countries in a fair manner, possibly based on market mechanisms and industry pledges. Friends of the Earth, however, appeared to advocate a more 'radical' or 'leftist' approach to the issue, disapproving of the use of market mechanisms and clearly linking climate change to the issue of global inequality. In this view, the responsibility for dealing with the issue lay clearly with the West and should not be shoved off onto developing countries. When it came to Cancún, this position was substantiated through demands for rather ambitious summit outcomes. In clear opposition to the 'even small steps will do' approach of mainstream NGOs, Friends of the Earth did not abandon its resolute rhetoric after Copenhagen and maintained it throughout Cancún.

Part of this rhetoric was an open letter to the Mexican government, which Friends of the Earth published at the beginning of the summit and accused the COP presidency of engaging in covert coordination and pre-negotiations with selected countries and thereby basically repeating the mistakes of Copenhagen. The letter did not resonate well with other NGOs, which actually approved of the

Mexican's leadership for the most part and feared that the move could be an obstacle to regaining momentum in the talks. Friends of the Earth's strategy of more pronounced confrontation did not match other NGOs' overall attitude.

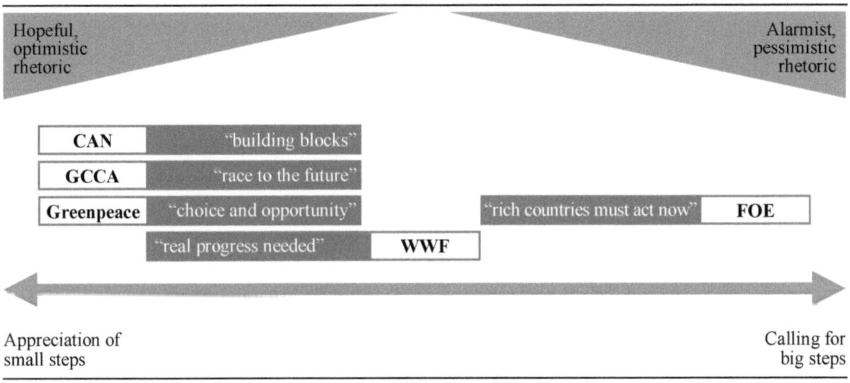

Figure 6.2: Spectrum of NGOs' messaging strategies

It was not like Friends of the Earth was without any doubts about this strategy. In the first half of 2010, in the aftermath of the failure of Copenhagen, the NGO had actually considered not giving much attention to the subsequent summit in Cancún and mostly ignoring it in its communication work. The frustration about the lack of concrete outcomes in Copenhagen was so strong that the suitability of COPs as a platform for initiating progress was fundamentally put into question. In this mood, even the sending of one communication professional to Cancún was seen as a waste of resources. But this changed in the second half of the year, when through the work of the Energy and Climate Justice working-group, Cancún came into focus nonetheless. FOEI's communications team started preparing for Cancún, mainly by crafting concise messages that put the demands of the working-group into media-friendly terms, supplemented by pre-worded answers to potential questions. This internal document was to serve as the overall guideline for all summit communication.

With this set of ambitious messages, Friends of the Earth seemed to distinguish itself from the 'building block' or 'step-by-step' frames of other NGOs (see figure 6.2 above), which deliberately lowered expectations toward the Cancún outcomes. This had somewhat paradoxical consequences: On one hand, Friends of the Earth already knew prior to the summit that the likelihood of seeing its high demands realized was close to zero and viewed itself as an isolated corrective that may only prevent worst-case results. On the other hand, after the summit, it still voiced disappointment with the lack of resonance of its message with other political actors and with the small political progress made at the summit in general. Essentially, the situation seemed to remind FOEI communication professionals a lot of the Copenhagen aftermath and caused them to express similar doubts about whether summit-based communication is the way to go forward and what role the NGO should play at COP-17 in Durban.

This frustration was strengthened by the fact that most other large NGOs – as part of their pragmatic, less harsh rhetoric – had welcomed the Cancún outcome or at least acknowledged its contribution to obtaining a full treaty at some point in the future. Hence, in rejecting the outcome, Friends of the Earth felt rather isolated (along with the government of Bolivia, with whom it had formed an alliance) – but still on the right side. The NGOs' communicators explained that they did not see why they should suddenly endorse outcomes in Cancún that they had still denounced in Copenhagen. According to them, even though expectations of what is realistic had changed, estimates of what is needed had not. Essentially, they wanted to measure political progress at COPs against absolute requirements of fighting climate change – and not just the relative opportunities provided by the negotiations. And so, whilst most other NGOs changed their strategy after Copenhagen and moved to more pragmatic, less ambitious approaches, Friends of the Earth refused to do so for political reasons. This essentially led to disappointment on both sides and reinforced the division between GCCA, CAN International, Greenpeace, and WWF with their reworked messaging on one side and Friends of the Earth promoting more ambitious, 'radical' messages on the other side.

6.2.4 Media relations activities

Differences between political actors' communication strategies could also be detected by observing the concrete communication activities that they conducted in pursuit of strategies. Working with journalists, i.e. doing media relations, was an essential component of these activities (other PR activities are dealt with in the following subsection) but came in different forms and intensities. Especially on the side of governments, notable variances were visible. Here, the most active players seemed to be Brazil, Mexico, and the US, which conducted media briefings on an almost daily basis and put in place structures and processes specifically aimed at easing the work of journalists. In the case of Brazil and Mexico, this seemed to be part of their general aspirations to attain international visibility (as was Mexico's hosting of the summit in the first place); the US, however, did not appear to connect any particular ambitions with their media relations but conducted them in a rather routine, subdued fashion.

Brazil, Mexico, and the US: Formal media relations with international outlook

The backbone of *Brazil*'s media relations was the daily media briefing at six p.m., usually given by one of the official spokespeople, Ambassador Figueiredo, Brazil's chief negotiator, or Ambassador Serra, special ambassador for climate change, in English or Spanish language. During the high-level segment in the second week of the summit, the Brazilian minister for environment, Izabella Teixeira, also appeared on the panel. The briefings were usually chaired by the head of SECOM's international branch. Further media relations activities included more informal briefings with selected journalists as well as scheduled interviews by the named spokespeople. Special emphasis was put on working with Anglo-Saxon high-quality press, such as *The New York Times*, *The Financial Times*, or *The Guardian*, illustrating Brazil's attempt to promote its image among decision-makers and academics – as opposed to a wider, more diffuse public. *The Guardian* was described as being particularly important, as its location of publication (London) was seen as the hub of activity for many ENGOs. In the Brazilian communication strategy, these organizations were understood to be a valuable avenue for talking to and influencing decision-makers. For the same reason, emphasis was also put on working with online and special-interest media, such as *The Huffington Post* or *ClimateWire*. This approach of selected target media was complemented by the usual interactions with larger national and international networks as well as news agencies.

As part of the media relations of *Mexico*, media briefings were held every morning, often featuring Mexico's chief negotiator, Luis Alfonso de Alba, as spokesperson on the panel. Press releases were made available on the website of the Mexican foreign affairs ministry. Due to a division of tasks with the UNFCCC (see 6.1.1.4), domestic media played the dominant role in daily media relations, with the Mexican TV network Televisa cited as prime target outlet. Internationally, *The New York Times*, *The Financial Times* and the BBC – along with major news agencies – were mentioned as the most important target outlets, although it remained unclear to what extent contact had been established during the summit.

In the case of the *United States*, formal media briefings were held almost on a daily basis, usually featuring chief negotiator Todd Stern or his deputy Jonathan Pershing on the dais. Transcripts of most briefings were published on the US delegation's dedicated COP-16 website afterwards. In addition, one background meeting was organized for a handful of selected media representatives.

One of the attending journalists saw two purposes in that meeting: selling spin and shedding light on technicalities:

> "They wanted to sell their spin, but they were also there to take our questions, to make sure that we understood some of the complicated technical things, we're understanding it from their point of view. But I think we all know that's one point of view, and if we go talk to China or Bolivia, we are going to find another point of view. At the same time, it was very helpful to have the delegation and their press folks there to answer some of the difficult stuff for us" (3843, p. 2, 25).

Outside formal media briefings or background meetings, Stern's spokesperson was described as being very responsive to ad-hoc media requests:

> "He really was wherever Todd was and then arranging… dealing with reporters who had questions. I found him to be very accessible, accessible mostly by Blackberry, but if I really needed to talk to him, it was very easy: he was always able to say 'alright, I will be in front of whatever, Luna room in 20 minutes' and be there and talk for a few minutes" (3834, p. 7, 29).

As described above, the US delegation appeared to consider media relations a routine element of summit business but did not use it for particular messaging beyond a commenting on the negotiations' progress.

India: Improvised yet effective networking

In clear contrast to this traditional approach to summit communication, the media relations of *India* constituted a special case, as they were run in a more informal, engaging manner. While there were sporadic media briefings (only one was formally announced in the daily program), most interaction with journalists seemed to flow more spontaneously. This was also facilitated by Ramesh's feeling comfortable being surrounded by press and maintaining continual contact, as his closest aide explained:

> "Our minister is very friendly with the press, in the sense that he believes in transparency. I would say two or three times a day, he would talk to the media contingent […] just to give an update on how things are going – to the extent that he could talk about it, because our approach to the media was that more is better and that transparency is important. If you don't tell people, there are conspiracy theories, there is misinformation, so we tried to be as transparent and as real-time as we could in our response to the media – directly through the minister and through the head staff" (6468, p. 4, 5).

Interviews seemed to be arranged by the minister himself or one of his closest staff, whose positions entailed not so much professional communication management but a much wider array of tasks. The absence of distinct communication staff could also be seen in the lack of a delegation-run mailing list, a fundamental component of most other delegations' PR. Instead, such infrastructure was created by one of the journalists, with the minister's top aide being added only later. However, he then used that list for "almost daily updates: both informal and formal press releases, when they were necessary" (6468, p. 3, 34). In the focus of India's international media relations were large news agencies as well as outlets of international reputation, namely "newspapers and publications that lead to, that develop a point of view which people read about and form their opinions on, for example *The Guardian*, *The New York Times*, *The Washington Post*" (6468, p. 4, 22).

Germany and South Africa: Low-profile interactions with domestic journalists

The media relations of Germany and South Africa appeared to keep the lowest profile of all countries in the sample. While the two other BRICS countries, Brazil and India, seemed to utilize the summit for media relations with an explicitly global orientation and the general ambition of becoming visible on the world stage, this did not apply to media relations of *South Africa*, the latest addition to the BRICS group. Instead, the South African delegation was mostly in touch with South African media outlets, some of which even received government subsidies for reporting on the significance of climate change and the respective international talks:

> "As part of our media strategy, we've also invested in bringing some of the South African journalists from home to here. We realized that because of the recession, a lot of media houses were not going to be able to pay for their media to come here. [But] this an important story to South Africa, this an important story to South Africans, and putting money into bringing journalists here is an investment, in the sense that the story of climate change negotiations is able to reach South Africans. The story of the threat of climate change is able to reach South Africans. And the story of South Africa's hosting of the climate change talks next year is able to reach South Africans" (1807, p. 3, 14).

This focus on South Africa was visible also in the choice of preferred target outlets, which did not comprise international media but seemed to feature domestic outlets in South Africa alone. Hence, the concrete media relations undertaken in Cancún appeared to be mostly geared towards these media and conducted in informal ways. The daily program did not list any formal media briefing by the South African delegation for the duration of the summit.

Similarly, *Germany*'s media relations seemed to be based primarily on individual interactions between German journalists and political staffers at delegation offices. Media briefings held by the German delegation were rare – formal meetings with journalists only took place when the German minister of the environment was present in Cancún for the high-level segment near the end of the summit. No media staff was present at the summit beforehand. When the minister arrived, he was accompanied by his chief spokesperson. During the summit, four press releases relating to the summit were published on the ministry's website, including English-language translations.

Governments' media relations hence came in the form of formal efforts with regular briefings (Brazil, Mexico, US), improvised but still effective networking (India), and low-profile, informal interactions with domestic journalists (South Africa, Germany). But also on the side of NGOs, approaches to working with journalists varied. Generally, the division between Friends of the Earth and other NGOs that became visible with regard to messaging strategy appeared to be replicated at the level of media relations. While CAN International's daily media briefing was supported by member NGOs and appeared to be the central stage for their press work, Friends of the Earth ran its own daily media briefing and did not seem to contribute to the alliances' media relations.

CAN International and GCCA: Coordinating media relations across NGOs

CAN International's briefings featured spokespersons from various member NGOs on the panel and were usually chaired by CAN International's communication consultant. The underlying idea of these briefings seemed to be to showcase the diversity of subject expertise and geographical spread present in the alliance, but also to speak with one unified voice. The briefings provided NGOs lacking own communication structures at Cancún with a stage for their positions but still linked them with CAN International. Briefings were supplemented by one-on-one interactions with individual journalists, giving preference mostly to international news agencies, which – according to CAN International's Director – "have some very, I think, quite bright and good reporters that follow these negotiations, so you can have a pretty substantial discussion with them" (3148, p. 7, 20).

In its function as behind-the-scenes coordinator, *GCCA* did not hold its own media briefings but contributed to the strategizing behind CAN International's briefings, for example regarding the staffing of panels, the coaching of speakers, or the selection of targeted outlets. In general, GCCA saw media outlets with international reach as primary targets of summit media relations, among them news agencies like Reuters, AP, and AFP and TV networks like BBC World News and Al-Jazeera English. However, GCCA's chief communicator also noted that with the exception of Al-Jazeera English, these outlets are mostly consumed in the West. Hence, when it comes to more specific messages that are aimed at particular markets (like in the case of the ad-hoc campaign against Japan – see section 7.2) or certain segments of the international audience, the selection of target media has to be more thought through. Outlets with large-scale reach beyond the West include Spanish news agency EFE for Latin America, Chinese news agency Xinhua, or Indian daily *The Times of India.* Alternatively, particular domestic outlets or specialized publications consumed by experts could also be targeted. Essentially, the selection of target outlets boils down to the question of "do I want global reach, do I want a certain region or country, or do I want a certain audience" (2135, p. 18, 10).

Greenpeace and WWF: Close one-on-one interactions with journalists

As member of CAN International, *Greenpeace International* commonly had a representative on the panel of the alliance's media briefings but also organized a separate briefing on December 7 (preceded by a related side-event on December 3) on the IT industry's role in building a clean-energy economy. In its media relations, the NGO generally targeted two audiences: negotiators on site and their respective leaders back in the capitals as well as supporters and people around the world sympathetic to climate action. For reaching both, international news agencies were seen as most effective channel:

> "Reuters, AP, AFP, Bloomberg, dpa, Point Carbon! [...] And their broadcasters: Reuters TV and APTN. It gets you global. If you get those guys, you get global coverage. [...] If we do nothing else, we do the wires!" (5356, p. 8, 12)

The primary focus in Greenpeace International's media relations, however, appeared to rest not so much on briefing several journalists at the same time but on individually working with them outside the press conference rooms, on assisting them in writing their stories and providing them with story proposals. This activity worked effectively, as it seemed, also because Greenpeace International was able to draw from its good reputation as information broker and provider of climate expertise with journalists (see de Jong, 2005; Vowe, 2006). However, it became clear in the course of doing media relations that some journalists remained skeptical towards the apparent change in messaging described above (see 6.2.2). Especially journalists interested in the nitty-gritty of political substance were irritated by the NGO's new rhetoric and also seemed to find abstract messages of 'choice' and 'opportunity' incompatible with their hard-news approach, as the NGO's chief communicator admitted:

> "One journalist came up to me at one point and said 'is there some sort of greenwash going on?', and I said 'what do you mean?', and he said 'well, everybody is just all happy-clappy about this stuff and how we can get somewhere – I don't think it's true'" (5337, p. 2, 20).

Hence, Greenpeace International's internal objective of its new messaging being picked up and distributed by journalists might not have been fulfilled with every media representative. Nonetheless, especially towards the end of the summit, there were indications that especially Greenpeace International had significant influence on how journalists reported on the summit (see 7.1.1 for details on the relationship between the NGO and journalists in Cancún).

Similar to Greenpeace International, *WWF International* was also part of the coordination behind CAN International's media relations but at the same time seemed to focus on its approach of "educating journalists" (6282, p. 1, 10), which in the words of WWF's chief communicator on site, was aimed at making the coverage of the negotiations "clearer, crisper, and more politically effective" (6280, p. 6, 25), while strengthening WWF International's position as provider of background and analysis for journalists. Two components were part of this particular approach to doing media relations: Firstly, background briefings for smaller groups of selected journalists played a major role. These events were set up to provide quick introductions to the state of negotiations from WWF's perspective and then to give journalists the chance of asking questions or going over particular aspects of the rather complex substance. This activity clearly drew on WWF's expert knowledge on climate issues:

> "Because we have such an extended and very competent policy team, our role in explaining detailed issues to media, so that they can get it quickly, is actually important. And that's something that others cannot play that easily. So that's the emphasis and that works" (6282, p. 3, 2).

Secondly, journalists were also informed by means of concise briefing papers, which were sent out daily under the heading of 'what's hot today in Cancún' to "point journalists to things that were going on" (6297, p. 1, 26). Additionally, special information sheets were provided on an irregular basis, which summarized on about two pages what a particular technical term or abbreviation relating to the negotiations meant:

> "[This was provided for] any journalist coming there and saying 'LULUCF, for God's sake, what is this'. And you can't find simple media information about it! So you just have these briefings. And we send them out when LULUCF becomes topical. So when in the morning we go out and say 'today, LULUCF is hot in Cancún', then we say 'come here or we will send you a brief about this in two hours, where you see what it is and why is it hot today and who can you talk to when you want to know more'" (6297, p. 7, 26).

Preferred target outlets for these activities were 'quality' media, i.e. media that provide some analysis of the talks. Here, reputable print publications, such as *TIME*, *The Financial Times*, or *The New York Times* were explicitly mentioned. More generally, news agencies Reuters and AP are considered channels of choice, as they – taken together – cover large parts of the globe. AFP, Xinhua, and EFE are regarded as second-tier agencies by WWF International, but play a very important role nonetheless. For background stories of a more technical nature, specialized publications such as *Nature* might be appropriate. The rationale behind identifying these outlets as preferred ones was to reach decision-makers in politics and business – the declared target group of WWF International's summit communication. Media fallout was monitored by a WWF staffer in Washington, D.C., who compiled daily summaries of how COP-16 was reported on in about 15 countries.

Friends of the Earth: media relations in cooperation with national chapters

As mentioned above, *Friends of the Earth* abstained from collaborating with the NGOs organized within CAN International and GCCA in the field of media relations. Instead, it conducted its own daily media briefing, which was commonly devoted to a specific topic, either an aspect of the negotiations or a study that the NGO had conducted. The composition of the panel was carefully chosen so as to avoid Western dominance and to represent the full range of countries with Friends of the Earth chapters. Generally, national chapters played a pronounced role in the NGO's media relations at the summit. While the federation's chair Nnimmo Bassey acted as international spokesperson and responded to broad media inquiries regarding the progress of the talks, more detailed questions were referred to specific experts in national member organizations. In the view of the international secretariat, it did not matter whether sources cited in summit coverage belonged to Friends of the Earth Spain, Friends of the Earth Middle East, or Friends of the Earth International – as long as the NGO was mentioned in one way or the other.

As with the other NGOs in the sample, media briefings – as well as individual interactions with journalists – were mainly directed towards international news agencies, including 'alternative' agency IPS, and TV networks, including Al-Jazeera English. The explicit inclusion of IPS and Al-Jazeera English may hint towards a preference for 'contra-flow' media (Thussu, 2007), which may be more open for Friends of the Earth's developing-world perspectives and global justice messages. In order to evaluate the success of media relations, specific targets were developed prior to the summit, such as 'ten stories in major global news sources', '40 stories in important national media' or 'no negative press'. The existence of such clear benchmarks points to some degree of professionalism with which media relations were conducted.

6.2.5 Public outreach activities

While media relations were directly targeting journalists, all political actors in my sample engaged in additional communication activities meant to influence the wider summit public. In the case of governments, such activities typically comprised the running of a country pavilion or information booth, the hosting of a side-event, or both. Generally, governments appeared to draw a clear line between these activities and their media relations. Both components of summit PR were commonly run by separate teams and not integrated with each other. A governmental delegation's media relations, for example, would rarely relate to the messages publicized or events held at its country pavilion.

Brazil, Mexico, and the US: Country pavilions of varying opulence

Out of all the pavilions and information booths operated by political actors at the summit, *Brazil*'s *Espaço Brasil* (see figure 6.3) stuck out for its meticulous design and integration of exhibits and multimedia. The theme of the pavilion was devoted to life in the Amazon, in line with Brazil's general message that the region required natural conservation but should not be excluded from economic progress (see 6.2.1). The facility featured an artificial walk-through rainforest – complete with bird sounds and rippling brooks – allowing visitors to get a feel for the scenery. On its dedicated summit website, the Brazilian government described the pavilion as "a sensory experience of the themes, colors, smells and sounds of the Amazon, […] a panel of cultural diversity in the local communities, of the forest peoples, and of the innovative projects, emphasizing local development solutions in the region which is home to over 26 million people" (Brasil.gov.br., n.d.).

Besides the exhibition, the pavilion also comprised a reception area and a lecture theatre for side-events. The entire facility was funded as part of a public-private partnership by both the Brazilian government and the private sector.

> "Most of the firms that are represented in our pavilion, in our exhibition, are companies that have some programs in the Amazon, that have some activity in the Amazon – and sustainable activity of course! So I believe it costs money for our taxpayers, sure, of course, but I believe it's worth it, because in the end, all the country, Brazil, will gain from this effort! We'll have a clear understanding of Brazil. And also we'd like to have more visits to Brazil as well, so everybody gains with that: the government and civil society" (6741, p. 10, 13).

Activities at the pavilion, such as side-events, were commonly highlighted in emails sent out to journalists by the Brazilian delegation. As a noteworthy exception, media relations and public outreach seemed to be well integrated in the case of Brazilian summit PR.

A similar approach was taken by *Mexico* with its *Pabellón de Mexico.* While clearly not as ornate as the Brazilian pavilion, it showcased some of the cultural heritage of Mexico. The integrated lecture theatre served as location for a rich program of side events ranging from talks on Mexican climate initiatives to film screenings. Brochures on Mexican climate action were made available on site and also posted on the official summit website that was run by the Mexican government (see figure 6.4). The third country in my sample operating a separate country pavilion was the *United States*. The 'U.S. Center 2010 Cancún' (see figure 6.5) was set up by the U.S. State Department and featured a room in which US research and actions relating to climate change were presented on multimedia screens as well as an auditorium housing several talks by diverse speakers, among them cabinet members. The schedule of side-events taking place in that auditorium was published in a separate brochure distributed around the Center as well as on a dedicated website.

In general, the US facility did not seem to convey a clear-cut message that was instantly noticeable when visiting. It did not appear to be integrated in a particular communication strategy but rather to serve as a platform for a diverse set of activities. It also did not convey a particular visual appeal or was designed in an inviting manner. Its container-like set-up allowed entrance only through two small doors, with the front room being only scarcely lit and sparsely furnished. Most notably, a reception area or front desk, which most other pavilions or information booths featured, was missing completely. In sum, it remained unclear what particular communication objective the Center was to fulfill.

Figure 6.3: Brazilian country pavilion, *Espaço Brasil* (author's collection)

Figure 6.4: Summit website run by Mexican host government (COP-16, n.d. a)

Figure 6.5: US country pavilion, U.S. Center 2010 Cancun (author's collection)

South Africa, Germany, and India: Smaller information booths

The other three countries in the sample operated smaller information booths, *South Africa* even two: At *Cancunmesse*, a small booth operated by the South African Tourism agency promoted the country as tourist destination and, through large-scale pictures and high-quality catalogues, contributed to nation-branding. The other facility at Moon Palace was run by the environment department in cooperation with the City of Durban and provided information on the infrastructure and logistics surrounding the subsequent COP in Durban. By having booths at both conference venues, South Africa was clearly visible for conference participants, although the delegation did not appear to host any additional side-events.

The information booth of *Germany* (see figure 6.6) was set up to be a general point of contact and platform for all summit participants interested in German initiatives and positions. In line with the broad messaging approach of the German delegation (see 6.2.1), the target group was described to be the wider environmental community present in Cancún. Visitors came from parties and from NGOs in roughly equal shares. Up to 100 concrete conversations were held at the booth as a result of such drop-bys, with the first summit week featuring significantly more visitors than the second. Among the visitors, those from developing countries seeking information about potential funding for domestic projects made up a large portion.

Figure 6.6: German information booth (author's collection)

What turned out to be a nice conversation starter at the booth was a box of small pellets for carbon-free heating, which visitors could pick-up and touch:

> "It's funny, but in many parts of the world, people want something that they can feel and touch. And then they can talk about, and so it's easy to open talk. There [are] few people – I think more the European ones, the Northern Europeans – they stand there and wouldn't even ask or dare to touch. But

> the others, they come in and 'oh, what's that, great, let me feel!'" (8282, p. 3, 21)

Finally, in the case of *India*, it was difficult to identify public outreach as separate component of the country's summit PR. While it had a specially designed information booth in place at COP-16, the facility was left largely unstaffed over the course of the summit. However, India conducted a side-event at *Cancunmesse* on December 4 titled 'Equitable access to carbon space: A paradigm for agreement', complemented by the release of a formal press release that was also posted on the ministry's website. Otherwise, the Indian delegation's visibility in terms of formal communication activities was limited.

While governments' public outreach mainly took place at their pavilions and booths, NGOs often moved to the alleyways and corridors of the summit venue in order to carry out communication activities supplementing their media relations. Most of these actions appeared to be in line with a tradition of inventive NGO campaigning and were hence clearly distinct from governments' efforts. One of such instruments was the staging of photo ops, or stunts. Other than that, NGOs engaged in more general protest activities, released publications specifically developed for the summit, and ran the usual information booths and side-events, which are not elaborated on here.

CAN International: The Fossil of the Day award show and *Eco*

One of NGOs' most visible communication activities was the Fossil of the Day award show that was put on every weekday evening at *Cancunmesse*. Organized by *CAN International*, 'the Fossil' has become an institution of its own at COPs, regularly drawing crowds of conference participants stopping for a look on their way out to the hotel. The awards are awarded to those players at a COP seen by CAN International as behaving in a particularly destructive, climate-threatening manner (Oberthür et al. 2002, p. 135). Pursuing an approach of public shaming, the Fossil is supposed to move particular players in the spotlight and condemn their demeanor in a fun, ironic way, according to CAN International's Director:

> "We have a discussion during the nominations about why a country might be getting a Fossil award, and it's important to have an understanding of what they did to deserve it and what the purpose of giving the award is, whether it's going to have an impact or not. So there may be cases where we see things that we really dislike but it actually isn't worth giving a Fossil award, too, because with scarcity comes more impact. And so if you give a Fossil award to a country that maybe deserves it, but it won't impact their negotiations, it might change the impact of those awards" (3148, p. 3, 17)

Hence, the selection of award 'winners' seems to be influenced by the likelihood of initiating policy change through that move. More stubborn, inflexible players apparently face a lower risk of being awarded than those known to be responsive to public pressure – even though this statement seems questionable when looking at the actual scoreboard (see table 6.1): In Cancún, most awards went to countries known for their rather inflexible positions: Canada (six awards), the US (five awards), Japan, Russia, and Saudi-Arabia (two awards each).

Date	First place	Second place	Third place
November 29	Canada	Canada	Canada
November 30	Japan	-	-
December 1	Saudi-Arabia, Norway, Kuwait, Algeria, UAE, Egypt, Iraq, Qatar, Jordan	Ukraine, Russia, New Zealand, Australia	-
December 2	-	-	-
December 3	Saudi-Arabia	-	-
December 4	Canada	-	-
December 6	-	-	-
December 7	US	Canada	EU
December 8	Japan	US	US
December 9	Canada	Papua New Guinea	US
December 10	US	Russia	Venezuela, Saudi-Arabia

Table 6.1: Recipients of CAN International's Fossil of the Day award over the course of the Cancún summit

The Fossil event is usually characterized by high entertainment value: Young campaigners play the roles of hosts and award recipients, integrating slapstick sessions and drawing on national clichés. The 10-minute show that usually takes place in front of CAN International's information booth is kicked off and concluded with a standard fanfare-like intro, which is usually sung live by some campaigners. The reasons for the choice of award recipients are then presented in a headline-like, humorous way. After the show, press releases detailing the accusations vis-à-vis awards recipients are handed out to media representatives present. The 'winners' are also announced in *ECO*, CAN International's summit newsletter, in the subsequent issue.

ECO can then also be counted as the third large communication activity by CAN International at the Cancún summit (besides media relations and the Fossil). In line with the policy focus of the alliance, it provides background on the subject matter of the negations as well as their progress (from an NGO perspective) but is also used for featuring smaller items of campaigning content from time to time (for example, see the 'contact ad' ridiculing Japan in section 7.2). While *ECO* came out every working day during the summit, the alliance also published a comprehensive once-off publication called *Cancún Building Blocks* at the summit. The brochure served as position paper, putting forward a proposal for what was to be achieved at the summit and carried the alliance's new modest messaging (see 6.2.2).

NGO protest and actions in Central Cancún

NGOs' public outreach also included protest activities outside the summit venue. In the course of the second week, *Friends of the Earth* – along with its allies *Via Campesina* and World March of Women – mobilized for a large-scale protest march through downtown Cancún and towards the conference grounds. Indicative of the rift in the NGO community (see 6.2.3), this mobilization ran independently from that of most other NGOs. Hence, two separate marches were held on December 7, 2010: one by Friends of the Earth and allies under the headline 'Thousands of Cancuns for climate justice!' and another one by the rest of the NGO community. While some communicators belonging to the two NGO alliances feared that this division could have been picked up by the media and used to illustrate a fundamental rift in the NGO community, it did not seem to have that effect.

More focused and concise than popular protest outside the confines of the summit venue are attention-grabbing photo ops, which particularly *GCCA* and *Greenpeace International* conducted at the Cancún summit. As a noteworthy feature of NGOs' summit PR, they are discussed separately in the next subsection (6.2.6). Interestingly, *WWF International*, which had stated that it would purposefully abstain from running any other communication activity than media relations, in one instance went against this line and contributed to a stunt-like activity. It was carried out in cooperation with WWF Mexico at the end of the first summit week at a shopping mall in Central Cancún. The activity involved a hundred youths in WWF T-shirts dancing to the sounds of a percussion group, shouting out demands, and eventually moving to the beach, where they used to candles to lay out the silhouette of the Americas. WWF's chief of summit communication recognized that this stunt might have been superficial on messaging but still considered it successful in its attraction of media attention. In general, he saw the careful tailoring of photo ops towards clear-cut messages as a prerequisite for their success. In Cancún, however, the limited amount of communication staff did not allow for this effort. Also, he diagnosed himself with some 'post-Copenhagen stunt fatigue', admitting that he might have had enough of them there.

Over the course of the summit, WWF International also worked on preparations for another, larger photo op involving a futuristic solar-powered boat that was to stop by at Cancún. The NGO tried to build a widely noted event around the docking of the vessel, preferably in the presence of, for example, high-ranking politicians of the Mexican government and the UN as well as the president of WWF International. The event was to highlight WWF's theme of green entrepreneurship and progressive technology being able to make a difference in combating climate change. It was to carry an optimistic scent of the future and was even thought to have the potential of becoming the closing visual of Cancún. While a lot of preparatory talks took place behind the scenes, the event could eventually not take place on the grand scale envisioned by WWF International due to logistical problems. When it became clear that the magnitude of the event would be significantly reduced as a result, the NGO dropped out of the preparations.

6.2.6 A closer look: NGOs' staging of photo opportunities

As mentioned earlier, one communication activity, for which especially Greenpeace possesses a special reputation, played a major role at the Cancún summit: photo ops, or stunts. The underlying idea of this activity is to induce media interest through the provision of amusing, creative, shocking, or otherwise attention-grabbing motifs for visual coverage. Ideally, these motifs convey a clear-cut, not overly complex message in an innovative way. Journalists are then invited to the scene, where their photographers or camera crews are given the opportunity of taking pictures or recording footage. Usually, the staging of a photo op is accompanied by a small media briefing, the opportunity of doing interviews with NGO representatives, or at least the distribution of a press release. Afterwards, NGOs commonly make their own images and footage available to news organizations – often through dedicated image databases on their websites or the picture desks of news agencies.

HIPS can be considered an effective environment for photo ops, since these meetings usually lack other visually appealing motifs that journalists could utilize for their coverage (beyond motorcades, handshakes, and heavily armed security personnel). Photo ops exploit this visually scarce environment. They close the gap between media's need for images and the lack of images at such events, as Greenpeace's head of summit PR explained:

> "Look, this is full of talking heads, the media go nuts! [...] And so the media are desperate for pictures. We know that. We know that, because we know that they have got their broadcasters come here with their cameras, just like 'getting pictures, getting pictures, getting pictures'" (5356, p. 7, 20).

Surrounding the Cancún summit, Greenpeace International organized or contributed to four photo ops. The first one took place on the day before the start of the summit, when the NGO flew a hot air balloon over the Mayan ruins at the world heritage site of Chichen Itza. The balloon, looking like a globe, carried banners reading 'rescue the climate' – a rather generic message – and 'make climate change history' – abstractly referring to the site's historic significance and the fact that, in contrast to past Mayan civilizations, modern civilization possesses the capacity of foreseeing climate threats and mitigating them (see figure 6.7). The timing of the stunt was chosen with a view to lighting conditions – and editorial deadlines, especially those of European newspapers, which were temporally closest to the event.

Figure 6.7: Hot-air balloon photo op by Greenpeace Int'l (Drieschner, 2010)

GCCA also provided media with a strong image at the beginning of the summit. It presented a 'Pyramid of Hope' – composed of cardboard boxes and covered with pictures and messages submitted by member NGOs – to UNFCCC Executive Secretary Christiana Figueres (see figure 6.8). As in Greenpeace International's Chichen Itza photo op, the idea was to offer to news outlets attractive visual material that they could use for illustrating their first stories on the summit. This way, chances of the summit's beginning figuring prominently in news media were enhanced. If a newspaper editor is given the opportunity of supplementing a story on the summit's beginning with a strong visual, he might turn it into a front page item.

Figure 6.8: Pyramid of Hope photo op by GCCA (Teriete, 2011)

In addition, Greenpeace International and GCCA conducted several smaller photo ops over the course of the summit: One of them incorporated life-sized human sculptures that were sunk eight meters deep to the sea floor and then, for the occasion of producing the images, visited by diving activists. Accompanying these underwater images, which Greenpeace had to distribute itself due to the difficult access for media, was the message that 'real people can't live under water'. In a similarly themed photo op, pictures of world-famous sights like the Eiffel Tower, the Statue of Liberty, or the Taj Mahal were pasted onto large boards and half submerged in the sea. This took place right at the beach, so that media could attend and produce their own images. For that purpose, favorable lighting conditions were again considered when setting the time of the stunt. According to Greenpeace representatives, both photo ops matched Greenpeace International's larger message of 'choice' in so far as they illustrated gloomy scenarios of the future that could be mitigated if humanity chose to do so.

GCCA, as part of an ad-hoc response to Japan's denunciation of the Kyoto Protocol, organized a photo opportunity titled 'Show some love to Kyoto', which involved an oversized heart with which conference participants could have a picture taken and thereby express their love for the Kyoto Protocol. This photo op was hence of interactive nature – it relied on ordinary participants' contributions – and emphasized the dimension of humor and ridicule that many photo ops draw on (see section 7.2 for details and an image).

Figure 6.9: Life-ring photo op by Greenpeace Int'l and GCCA (Teriete, 2011)

Most successful in terms of media fallout appeared to be the fourth and final photo op, which Greenpeace International staged together with GCCA on the morning of the final summit day. A giant, bright orange life ring was brought to the beach, where it was supposed to rescue negotiators – campaigners dressed in suits – in danger of drowning in the sea. Later, the same campaigners formed the phrase 'hope?' next to the ring (see figure 6.9). While this photo op was originally not conceived to create the final visual coming out of Cancún, it resonated well with media, as GCCA's Communications Director pointed out in his description of the stunt:

> "On the Friday morning, a [large, 15 square meter] life ring was brought to a beach, and negotiators were swimming in the water in suits – people dressed as negotiators – and we threw them a lifeline. [...] And that was also videotaped from a helicopter. We did these aerial photos, and [...] after the life ring was back on the beach, people grouped next to it to form a big 'hope question mark'. And that is the picture that got a lot of pick-up. And we didn't intend it to be a closing picture, we didn't think it would be the closing image or closing message, we thought it was our '24 hours to go message', but it turned out to be the closing image, because somehow Cancún ended on this mixed note – somehow won, but not really won much – so this lifeline and 'hope question mark' message seemed to fit for media. So many used this to illustrate their closing articles from Cancún, and I thought that was a very popular image and successful in terms of getting coverage" (2176, p. 7, 4).

The nature of this photo op as joint operation between GCCA and Greenpeace was also apparent from both logos (TckTckTck and Greenpeace) being displayed on the life ring. According to GCCA, Greenpeace, as the logistics leader, took care of obtaining the life ring, moving it to the beach, and recruiting some of the volunteers forming the message. Aerial photography and videography, on the other hand, was carried out by GCCA. This example illustrates how the provision of support by GCCA to its member NGOs may work, although Greenpeace, as a resourceful and professionalized player, might require a different kind of assistance than smaller players with less pronounced communication capacity at these events.

Greenpeace's general photo op strategy

Greenpeace International was certainly satisfied with the extent to which media drew on images of its activities, especially in the beginning and at the end of the summit:

> "We aim for the opening and the end of it, because you usually get some global attention in those two moments. The balloon and the life ring achieved that. If anything went global, it was our pictures!" (5337, p. 9, 30)

> "The images all flew around the world. We've don't often get front page *Washington Post* and front page *LA Times* in the same week!" (5337, p. 1, 13)

According to own descriptions, the NGO tries to adhere to some basic principles when crafting photo ops of worldwide reach. For example, it is made sure that the motif also works across cultural borders. As part of this effort, sights and icons that are known around the world have become a popular vehicle for Greenpeace's photo ops, such as the Mayan pyramids for the balloon stunt or the archetypal Caribbean beach – complete with white sand and crystal-clear water – for the life ring stunt. Through this approach, audiences are to integrate the visual with previous knowledge they possess (even if it is only a cliché they hold of a particular world region) and thereby pay closer attention to the message. According to the Media Coordinator, that message should ideally be transported without the usage of written language, e.g. on banners. If banners need to be involved, the written language used should be as simple and headline-like as possible. English words like 'hope' or 'action' might be suitable.

Beyond those principles regarding international comprehensibility, a number of more formal rules have also emerged over time (although for some of them, there seems to remain minor disagreement within Greenpeace): The significance of suitable lighting conditions (which sometimes occur at rather inconvenient times, such as early in the morning) was already pointed out above. Furthermore, if journalists are invited to the scene of a stunt to autonomously take images (as opposed to being supplied with NGO material), the site should be near, if not even directly at, the summit venue. Building a messaging connection between a summit and a far-off stunt is difficult; also, journalists cannot be expected to endure long journeys just to reach the site of a photo op. And lastly, classic Greenpeace motifs involving chimneys or power plants should be treated with care, as they might have been used too often.

6.3 The influence of long-term factors on strategy choice

Based on the explorations of political actors' general communication structures and processes (section 6.1) and their communication strategies and corresponding activities at COP-16 (section 6.2), I now review the factors that, as part of my conceptual model, were hypothesized to structurally shape actors' summit communication strategies: (1) an actor's state or non-state nature, (2) its worldviews and standpoints in the climate change debate, (3) the degree of professionalization in its political PR in general, and (4) the level of funding provided for summit communication. My findings on these four factors as well as their influence on communication strategies are laid out in this section. As described above (see 3.3.2), the factors were deduced from the literature, and the objective of fieldwork was to empirically test the validity of their assumed influence, to check up on their applicability to real-world summit communication. It became clear in the course of empirical analysis that the initial set of factors had to undergo some reworking; in three instances, the factors or their presumed influence could not be straightforwardly confirmed but had to be partially reconceptualized or even preplaced altogether. All modifications are elaborated on below when discussing the individual factors. The entire conceptual model in a revised form is presented in the conclusion (see section 8.1). Generally, the modified set of factors can be regarded as representing some significant – but by no means all – influences on the choice of communication strategy at HIPS. They should not be seen as suggesting simple monocausal explanations; the factors are thought to work in parallel and in varying constellations.

6.3.1 Type of actor

First and foremost, the way in which a political actor made use of the Cancún summit for communicative purposes depended on its state or non-state character. Governments and NGOs differed starkly in their general PR approaches, specifically in the clarity of strategy (which was usually higher for NGOs), the repertoire of concrete activities (which was usually broader for NGOs), and the efforts invested in evaluating summit communication (which was usually stronger for NGOs). As was already posited in the literature, the state/non-state dichotomy generally bears strong effects on an organization's communicative behavior, and this can be confirmed with regard to their communication strategies at the Cancún summit:

Government delegations participate in the summit primarily to make policy; for them, this assembly of officials from around the world is about reducing the transaction costs connected to negotiating international political agreements. While they might also have objectives relating to their international image and public standing (see Brazil, for example), these *communicative* objectives are mostly subordinate to and in support of *policy* objectives. This means that government delegations are not PR players per se – major parts of their work at a climate summit are actually about the opposite: to meet with counterparts behind closed doors (i.e. deliberately not in public) to discuss the detailed substance of climate policy (that is also not always well-suited for public communication). Governments' summit communication seems to constitute a secondary layer to a primary core of policy-making; doing PR is not their principal mission.

This looks different for NGOs. While many of them (some more than others) are very active in lobbying negotiators and thereby taking part in the non-public policy-making business behind the scenes, their non-institutionalized position at the negotiations forces them to routinely resort to PR. Working by means of public communication (for generating pressure, mobilizing protest, shaming politicians etc.) is their only lever; for some NGOs, like GCCA, running PR campaigns is all they do and essentially the purpose of their existence. As the communicative dimension is just more key to their work, NGOs naturally possess more experience and sophistication in managing public communication than state counterparts. It is unsurprising then that they outperformed governments in terms of clarity of strategy, repertoire of concrete activities, and efforts invested in evaluating summit communication.

It might not be unlikely that communication approaches of state and non-state actors will converge in the future and that governments might adopt communication activities which currently still seem to be the nearly exclusive terrain of NGOs, such as staging photo ops or utilizing visual trademarks. Some countries have taken first attempts at such tactics, e.g. the Maldives by staging an underwater cabinet meeting prior to the Copenhagen summit (Eide and Kunelius, 2010, pp. 11, 45) or Brazil by presenting a coherent 'nation brand' in Cancún. If such developments intensify, it might reduce the explanatory power of this particular factor for the choice of summit communication strategy.

6.3.2 Worldviews and standpoints

A second long-term factor that was confirmed as carrying explanatory value was the actor's overall worldviews or standpoints. While in the factor's original conception, this referred to actors' more specific positions *in the climate change debate*, this was adjusted to account for actors' more general stances *on the international stage*. Especially in the case of government delegations (but, to a smaller extent, also of NGOs), the particular communication approach carried out in Cancún seemed to embody and reflect the respective actors' general standing and ambitions on the world stage. Governments' public diplomacy at HIPS might be shaped by how they perceive hierarchies in the international system and which respective changes they strive for.

This was particularly visible for the BRICS countries, especially for Brazil and India, to a lesser extent also for South Africa (Russia and China were not included in my sample). Brazil and India both engaged in highly visible and self-conscious summit communication, even though by drastically different means. It seems sensible to attribute this grand appearance at the summit to both countries' general ambition to play a larger role on the international stage. While India gained visibility mainly through the stature and vigor of its environment minister, who ran the country's PR show on his own, without the help of dedicated communicators, Brazil followed a different path and had put in place a highly professionalized communication apparatus for the management of its international PR, not only at climate summits.

Brazil has openly acknowledged the connection between its concerted public diplomacy effort and its rise on the world stage. In a *Xinhua* news piece, the international director of Brazil's government communication office SECOM (who was also interviewed for this study) identified the promotion of the country's new international standing as the prime objective of his communicative work:

> "We show the idea of a country that has matured, which is consolidated as a strong and vibrant democracy, with problems and challenges, but that today presents itself to the world as a country with all conditions for an even bigger leap, strengthening its economy, reducing social inequality and investing in more advanced economic sectors" (Costa, 2010, ¶ 5).

A *Huffington Post* article confirmed the existence of this explicit approach – albeit from a more critical angle:

> "Brazilian authorities, however, are concerned that outsiders may ignore such advances and focus instead upon their country's horrific drug-related crime scene, urban *favelas* and startling rural injustice. In an effort to burnish Brazil's image, [former Brazilian president] Lula told SECOM, the president's own Secretariat for Social Communication, to focus laser-like on international public relations. In tandem with Lula's desires, SECOM has sought to give Brazil a makeover by emphasizing the country's solid democracy, robust economy, pacifist-oriented diplomacy and environmentally sustainable policies. SECOM directs its efforts at journalists, opinion makers, investors, academics and students while conducting key seminars, trips and interviews" (Kozloff, 2010, ¶ 3).

While Brazilian and Indian summit communication seemed to clearly reflect both countries' ambitions, a similar argument can be made for the PR of the more established players in my sample, the US and Germany. As mentioned throughout this chapter, both countries' communication appeared to lack a certain amount of inspiration, which could have been visible in a more pronounced strategic focus, more innovative messaging, or more visible activities. Instead, the summit PR of both parties came along in a rather traditional, unmotivated fashion, as if they had been doing it in similar ways for decades (which some participants posited). In my argument, this rather tranquil communication approach reflects clearly both countries' recognized positions as influential industrialized countries of the West, which also did not pursue ambitious foreign or international trade policy projects at the time of the summit. Both parties seemed to be rather content with their overall international standing – and their summit communication appeared to reflect just that. Hence, the worldviews and standpoints on the international stage might carry explanatory value even if they appear less pronounced or less infused with ambition. It has to be noted, however, that governments' worldviews and standpoints can change; the set of delegations most active in PR might look different from one climate summit to the next.

Differing political positions and ambitions could also be found within the NGO community, where organizations put forward diverse approaches for how to deal with the challenge of climate change. A particular rift could be seen between more progressive or 'radical' NGOs, which also employed rather harsh rhetoric in voicing their demands and evaluating the Cancún outcomes, and those taking a mainstream, sometimes business-friendly, approach that also became apparent in a new optimistic messaging strategy. As demonstrated throughout the chapter, these worldviews, or ideologies, did indeed shape NGOs' communication approaches at Cancún and lent them distinctive profiles – be it Friends of the Earth's cooperation with the *Via Campesina* movement and the leftist government of Bolivia, Greenpeace International's production of globally popular, aesthetic imagery, or WWF International's 'education' of journalists' and promotion of the business opportunities provided by the climate threat. NGOs come from different backgrounds and employ different worldviews; their summit communication is then shaped accordingly.

6.3.3 Degree of professionalization

The third long-term factor found to shape political actors' strategies was the degree of professionalization in actors' general communication structures and processes, also outside the summit. As described earlier (see 3.3.2), 'professionalization' here refers to the usage of specialized knowledge in communication processes, the integration of specialists with respective training or experience in communication structures, and the amount of resources deployed for the systematic preparation, implementation, and evaluation of PR. In its purely deductive, not yet empirically tested form, my conceptual model also featured 'degree of funding' as a separate factor hypothesized to shape strategy choice, but this was eventually merged into the 'degree of professionalization' factor, since the degree of professionalization and funding seemed to be confounded in reality.

On the whole, the NGOs in my sample were found to possess most of the professionalized PR capacity at the Cancún summit. All five organizations operated dedicated communication teams on site, which comprised professionals holding appropriate experience or training (such as a degree in journalism) and sometimes additional specialized competencies in particular PR fields (such as social media, photography, or graphic design). With minor exceptions, NGOs commonly infused their processes with current knowledge on how to do effective PR, for example by developing strategic frameworks, turning those into concrete messages and talking points, setting benchmarks for use in subsequent evaluations, fixing clear protocols and processes for summit business, and comprehensively evaluating their work afterwards. Based on NGO communicators' descriptions and my own observations of these processes, I could detect a rather high degree of professionalization for these organizations. They seemed to possess a clear understanding of the workings of international political PR as well as comprehensive experience in carrying it out. This is only stressed by the fact that two of the NGOs had their entire summit communication run by external communication consultants, who make a living by selling their PR expertise. However, one communicator pointed out that these comprehensive and professionalized communication structures are limited to a small number of large international NGOs and should not be taken for granted for smaller civil society actors:

> "I got the strong impression that you have something like Greenpeace, WWF, Oxfam, Friends of the Earth – and then a long time nothing – and then loads of small groups with hardly any media capacity, the capacity to do professional media and communications work. They have these people in their headquarters or in their offices back home, but they don't necessarily seem to bring them. So maybe they just have two or three people on the ground and these people might be, I don't know, advocates, lobbyists, activists, or something like this" (9876, p. 11, 21).

Things looked different on the side of government delegations. While all investigated countries (excluding India) had dedicated communication professionals on site (at least for some days), the carrying out of what would be considered professionalized PR processes could not be detected for all. For several actors, the overall strategy and corresponding messages were fuzzy or missing altogether, target groups and target outlets seemed not well-defined, and post-hoc evaluation was superficial and not based on preset benchmarks. The most notable exception to this was – unsurprisingly – Brazil, whose contracting of two PR firms for the management of summit communication was remarkable.

It has to be noted, though, that the lack of professionalized PR structures and processes in the case of some government delegations did not mean that their communicators were ineffective in what they were doing. Especially in terms of media relations, government spokespeople exercised significant skills in handling media requests and moderating press conferences. Similarly, public outreach staffers at country pavilions and information booths were welcoming and active and – like their media relations colleagues – able to reflect on their work. Above all, the case of the Indian delegation, which lacked professionalized communication structures and processes altogether and yet attained high degrees of visibility and appreciation, demonstrates that strong professionalization is not a prerequisite for effective summit PR and, as factor of influence, should not be overrated. Apparently, it can be 'overruled' by such factors as the ambitions on the international stage embodied by the Indian environment minister.

This does not mean that actors' strengths and deficiencies regarding professionalized PR do not become visible in concrete communication efforts, as high message discipline in press releases (indicating strengths) or unstaffed information booths (indicating deficiencies) illustrate. Especially governments' communication ignored major concepts and techniques of professionalized PR, like the combination of different activities (i.e. media relations and public outreach) into 'integrated communication', the definition of target groups, or the clarification of the overall communicative objective. With the exception of Brazil, all delegations showed room for professionalization in this regard. Nonetheless, the overall orientation of communication strategies appears to be more strongly shaped by other factors.

6.3.4 Individual background of PR professionals

The fourth long-term factor shaping political actors' strategies – the individual background of PR professionals – was detected in the course of fieldwork; it was not part of the initial version of the conceptual model. However, inclusion in the model appeared necessary, as I came across several instances of strategic decisions being motivated by communicators' individual competencies or preferences. While the type of actor, standpoints and worldviews, and the degree of professionalization might all set structural parameters for summit PR, significant influence also originates from the personal feelings, thoughts, and actions of communication professionals.

For example, WWF International's hesitation to conduct photo ops, which also seemed to have an effect on the overall amount of coverage initiated by WWF, was *partially* due to the fact that the communicator in charge was "somehow bored with it" (6282, p. 3, 1) and instead had the feeling that intensifying media relations was the way to go forward. Additionally, communicators' academic backgrounds appeared to have some – if minor – influence on the general approach they were taking in running summit communication. For example, having studied broadcast journalism, the communicator of the South African government seemed to focus a lot on broadcast media in his discussion of media relations. Similarly, the GGCA communicator, as a former student of communication studies, was able to provide some theoretical reflections on his strategic work at the summit. While these are all just smaller facets of individual backgrounds and we may only have a hunch regarding the clear nature of their influence, this is just to say that individual micro influence should also be seen as one of the factors shaping the choice of communication strategy at a summit.

6.4 A typology of summit PR approaches

All findings regarding the differences and particularities in state and non-state actors' communication structures and processes (section 6.1), their communication strategies and corresponding activities (section 6.2), and the influence of long-term, structural factors on these strategies (section 6.3) were synthesized for the purpose of building a typology of archetypal approaches of summit PR. The basis of this typifying was a thorough analysis of how actors' summit PR turned out on the various aspects of interest, or categories, inherent in my conceptual model. A comparative overview of some of these categories is provided by tables 6.2 and 6.3. If similar manifestations on a given category could be found across actors, they were clustered into groups of shared manifestation. In the end, it was checked whether some actors were bound by more groups of shared manifestation than others (see 4.5.2.2). Such actors seemed to represent a distinct approach to carrying out summit PR.

Governments showed three distinct approaches to summit PR at COP-16 (see figure 6.10):

- *Traditional, quiet media work:* This approach refers mainly to the US and Germany, which focused on formal media briefings and/or informal media contacts, supported by some low-key outreach work at an information booth/country pavilion. An overall strategic objective and corresponding messages of particular clarity could not be detected.
- *Professionalized, highly visible PR mix:* This approach refers mainly to Brazil, which carried out assertive media relations managed by external PR consultants as well as comprehensive public outreach activities at its elaborate country pavilion. Brazil's messaging reflected an ambitious strategic agenda.

The way in which Mexico and South Africa conducted summit PR shared characteristics with both aforementioned approaches and hence occupies a middle position. A separate archetypal form of doing summit PR was detected for the case of India:

- *Colorful one-man show:* India made up for its overall lack of communication structures through the efforts of its charismatic environment minister and gifted PR player. Despite the more free-flowing, improvised nature of its summit communication, a strategic agenda was detectible.

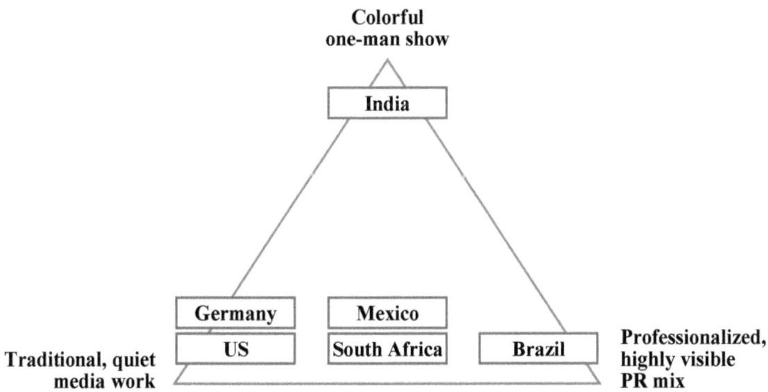

Figure 6.10: Typology of governments' summit PR approaches

NGOs, on the other hand, resorted to two different approaches (see figure 6.11):

- *Inventive campaigning:* This approach refers to CAN International, FOE International, and GCCA (although to varying degrees), which in addition to standard media relations, invested large resources in staging photo ops and other campaigning actions. Behind their work loomed a professionalized understanding of PR and clear-cut strategic objectives, visible, e.g., in thought-through messaging.
- *Educating press:* This approach refers mainly to WWF International, which deliberately abstained from engaging in campaigning actions (although contributory and conceptual efforts were taken) and instead focused on intimately working with journalists and providing them with analysis and background to the negotiations.

Greenpeace International should be placed between the two aforementioned approaches, since it combined both approaches in its summit PR.

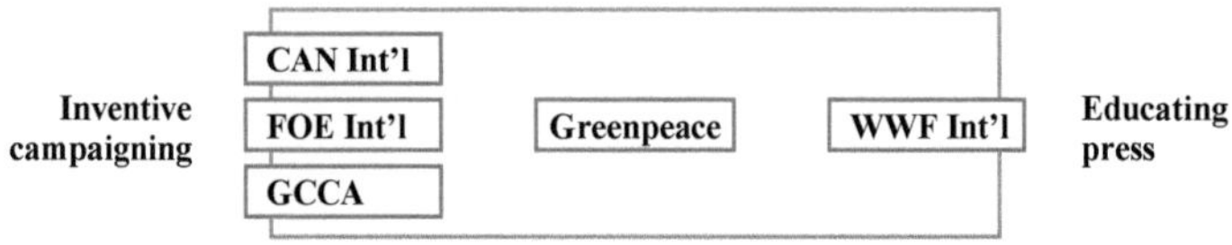

Figure 6.11: Typology of NGOs' summit PR approaches

6.5 Chapter summary

This chapter dealt with the structural, preplanned side of public diplomacy at HIPS. In the focus were, first, government and NGO delegations' communication structures and processes, which I described in some detail. Here, notable differences emerged, e.g., with regard to the amount of resources made available for communication purposes or the incorporation of external communication expertise. I then compared selected aspects regarding actors' strategic communication, namely their overall messaging approaches, ways of conducting media relations, and repertoires of additional public outreach activities. While governments' messaging is diverse and ranges from specific messages related to the negotiations to more general image messages, NGO messaging is divided into two approaches: Most NGOs went for a strategy of explicit expectation management, in which they avoided the instigation of high hopes and showed appreciation for less ambitious outcomes. The underlying motive was to avoid a symbolic overburdening of the summit (as at the 2009 summit in Copenhagen). Friends of the Earth stuck to its established approach of voicing rather ambitions demands.

Beyond messaging, actors' summit communication included several noteworthy features, like Brazil's opulent country pavilion, India's effective improvising, or the striking photo ops by GCCA and Greenpeace. Based on all findings, a typology of five approaches in summit PR was built: (1) *traditional, quiet media work*, (2) *professionalized, highly visible PR mix*, and (3) *colorful one-man show* on the side of governments as well as (1) *inventive campaigning* and (2) *educating press* on the side of NGOs. By and large, actors' approaches in summit PR are strongly shaped by their standpoints and ambitions on the international stage as well as, simply, their nature as state or non-state actor. Professionalization and PR professionals' individual backgrounds also pose an influence.

7. Short-term interactions between summit PR and summit proceedings

As argued in the theoretical part (see sections 3.2 and 3.3), HIPS, in this research, are conceived as spheres of resonance and feedback, engaging various actors in frequent, intensive interaction. Such events constitute focal points, not just geographically and temporally, but, above all, politically and communicatively. Actors respond to the exceptionality of such events by making exceptional amounts of resources available. They focus their efforts on being heard and achieving visibility – and on observing others' efforts in being heard and achieving visibility. What emerges is an environment of mutual monitoring and rapid response: Actors' communication activities are embedded in a flow of actions, reactions, and counter-reactions, of positioning and repositioning at the transnational level.

For this reason, when investigating political PR as it is being conducted at a climate summit, one should not only look at actors' preset PR strategies and activities (see chapter 6), but also at the consequences they bring about. These *short-term* effects at the event itself are discussed in the following two sections: How do other summit actors – particularly journalists, one of summit PR's prime target groups – process political actors' communicative offerings? And by which mechanisms does journalistic treatment of summit PR impact the summit itself and lead to further reactions (section 7.1)? Do short-term discursive opportunity structures emerge that cause political actors to adjust their communication strategies ad hoc (section 7.2)? On the other hand, political actors might also be subject to more *long-term* learning effects and draw lessons from the specific unfolding of the event. This constitutes a second feedback cycle, which is discussed in section 7.3. This chapter hence focuses on the procedural, dynamic, unpredictable side of HIPS.

7.1 Short-term effects of political actors' PR at the summit

As hypothesized in my conceptual model, the more short-term, on-site consequences of political actors' communication strategies might be visible in their treatment by journalists and in the impact their coverage has on summit proceedings. Firstly, I explore the noteworthy relationship between political PR and journalism at the summit, which – as I argue – saw PR professionals and journalists cooperating closely (7.1.1). Afterwards, I discuss how political PR professionals saw the short-term effects of summit coverage, to which they contributed through PR, on summit proceedings (7.1.2).

7.1.1 The relationship between PR professionals and journalists

It is my argument here that certain contextual parameters of the Cancún summit facilitated the emergence of a remarkable constellation between political PR professionals and journalists that essentially dissolved traditional boundaries between both sides and challenged their typical distribution of tasks. This constellation might be best described as networks of 'coproduction', in which political PR professionals and journalists did not hesitate to work hand in hand and cooperatively construct the worldwide image of the summit.

The prime factor facilitating the emergence of these networks of coproduction was the temporal and spatial confinement of the summit (see section 5.2) as well as the compression and acceleration of working habits that occur during such events. At COP-16, negotiators, NGO observers, and media representatives spent most of their workdays on an area of barely 500 x 500 meters. While the whole of the conference grounds covered a much wider area at two separate venues, the hub of all conference proceedings was a set of palm-lined pathways between the *Azteca* and *Maya* conference buildings at Moon Palace (see figure 5.2). This is where the action took place, where climate change professionals from around the world, both of political and journalistic nature, kept on running into each other for the duration of two weeks. They almost had no other choice than to interact and to network, to share opinions and to develop personal ties. A UN official familiar with the working habits of PR professionals and journalists summits described the atmosphere as 'camp feeling', with both sides being pulled out of their usual routines and placed in close proximity of their counterparts. In this specific setting, three distinct mechanisms conducive to the coproduction of summit coverage by political PR professionals and journalists could be detected.

7.1.1.1 NGO communicators as experts and information brokers

Firstly, NGOs' PR professionals were occasionally asked by journalists if they could shed light on some of the more intricate details of the negotiations or, more generally, if they had ideas for stories about the summit. NGO representatives would then provide journalists with input (general guidance as well as, for example, reports about NGO initiatives or data from NGO studies) or connect them with knowledgeable experts within the organization. Journalists' requests for NGO support even came in the form of humorous but yet serious 'give me a story!' cries, which an NGO's PR professional recalled in the interviews. Significantly, NGOs' evaluations of the summit communiqué, written up and distributed by NGOs' communicators in the early morning hours of the final summit night, were greeted by journalists with strong interest and comprehensive integration into their coverage. One NGO's communication manager was surprised herself about the extent of desired media fallout triggered by the release of these analyses:

> "We put out to the journalists a two to three page analysis of the key bits of the [final communiqué's] text: what was good, what was bad. [...] And we did that within a couple of hours of its happening, and I have never had so many journalists coming back to me and absolutely thanking me for what we did! [...] They don't have time, they don't have the expertise, half of them, to go through – some of them do – 25 pages of text and work out what's changed. They don't know it to that extent; they rely on us to do that! [...] I never had so many journalists thanking me" (0417, p. 13, 17).

With summit proceedings presenting themselves as rather obscure and complex to some journalists, relying on NGOs' expertise seemed to be the only option. This is also reflected in Boykoff's (2012) analysis of Cancún summit coverage, in which he identifies NGO officials as dominant suppliers of information and commentary (p. 255). But also governmental communicators seemed to be popular points of contact for journalists. The former would then suggest a perspective on the progress on the negotiations or put the inquiring journalist in touch with appropriate interviewees in the delegation. Working with governmental sources is of course usual procedure in political journalism, also outside of summits, but seen against the backdrop of unusually complex subject matter, high time pressure, and limited personnel, it is not unlikely that it also shapes coverage to an unusually high degree.

Significant in this example is the extent to which journalists surrender control of their coverage to PR professionals. The latter do not just make content offerings from which journalists choose independently but become 'information brokers' and analysts who are actively engaged in the construction of coverage. In particular, journalists' openness towards PR support meant that NGOs' new messaging strategy (see 6.2.2), which they concertedly rolled out in Cancún and also worked into their final analysis of summit outcomes, found particular resonance in media coverage. Apparently, NGOs can effectively set a tone in summit coverage if their concerted (re)framing efforts meet a setting of coproduction.

My argument is not that journalists completely drop their professional skepticism vis-à-vis government and NGO material. As a matter of fact, one NGO communicator described a small episode demonstrating the hesitation that journalists showed towards evaluations by NGOs and their new, positive rhetoric:

> "I had a very interesting discussion with [a correspondent from] the *Guardian* […]: This was Friday mid-morning, she was on deadline for Saturday's paper […]. And I was sitting opposite her, and she'd just come out of Connie's [= EU Climate Commissioner Connie Hedegaard] press conference, which was an EU press conference – it was like 'if we don't get anything, we're all destined for a dim, dark future', it was a very negative press conference! But Connie did that to kick some of the countries and to try and kick some life into the final few hours of the negotiations: a very dire, very warning type of press conference. But [the correspondent] was taking this as the final outcome! And she was writing her final story along those lines! And I had to say to her 'look, you do not know, anything can happen, it's all open, right now' – because it was! (0417, p. 6, 15)

But the journalist exercised caution at first:

> "And she just looked at me, and she decided, she said 'okay, I hear what you are saying, but I'm not going to go with it'. I said 'just be careful', I said, 'really, I wouldn't call the final outcome now and certainly not in such negative terms'. […] I was just really warning her!" (0417, p. 6, 23)

However, according to the NGO communicator, the journalist eventually bought into the frame and adjusted her coverage accordingly:

> "And she of course scrabbled to change the final online story, because the final outcome in the print *Guardian* was appalling and didn't match what actually happened […]. That was quite interesting. So we were walking around, giving that – not in sound bites – but just 'look guys, anything can happen'. And then we definitely took that attitude throughout the meeting, and that actually was quite interesting, because it made for relatively positive coverage at the end!" (0417, p. 6, 23)

Clearly, it would be wrong to assume that journalists could be spoon-fed by PR professionals. However, especially for journalists with less specialized knowledge of the negotiations, the analysis and guidance by spokespersons of prominent NGOs and government delegations were much appreciated.

7.1.1.2 Collaborative composing of visual coverage

The second mechanism can be seen in NGOs' carrying out of photo ops, or stunts (see 6.2.6), that provided journalists with those attention-grabbing images that they badly needed for their coverage of an otherwise visually not very exciting summit. Images of staged happenings would indeed dominate worldwide coverage of the summit, especially when they were distributed by one of the large news agencies. But what is remarkable here is not so much the eager use of those opportunities for reporting purposes but journalists' roles in constructing the motifs – and, through that, messages – of these stunts. For example, at the site of stunts, journalists and photographers would voice clear preferences vis-à-vis NGO representatives for how they wanted the performance to look, what kind of visual they would need, and how those involved should behave so as to increase media suitability for reporting purposes.

The insight of photo opportunities being an effective PR instrument has been long established in communication studies (Delicath and DeLuca, 2003; DeLuca, 2001; Greenberg, 1985). However, the role that journalists themselves play in composing these types of contents is noteworthy. At Cancún, it did not always seem like journalists would calmly screen what NGOs had to offer in terms of visual material and then make up their mind about the utility of these images. Instead, they would at times actively coproduce these stunts by giving NGO representatives concrete advice on what to do. While journalists might be motivated by the prospect of increasing the appeal of their coverage through striking images, they seem to forget that through their actions, they strengthen the effectiveness of what might rather deserve more detached observation.

7.1.1.3 Informal interactions in shared workspace

Thirdly, the close ties between NGOs' PR professionals and journalists could also be seen in the constant mutter and low-key interactions between both sides. This was often facilitated by the fact that some journalists perceived the official media center as inconveniently located (see 5.2.3) and, hence, opted for working from general working zones, where they would be sitting right next to NGOs' PR professionals (sometimes literally less than a meter apart from each other), as confirmed by the UN communicator:

> "In front of the press conference room, you had tables, and the tables were shared by media and by NGOs. So they were very close – for the full two weeks at least!" (3321, p. 6, 19)

This setting allowed for several episodes of spontaneous, informal collaboration. One NGO, for example, pragmatically initiated the revision of a picture caption that an international news agency had distributed alongside an image of one of the organization's protest stunts. NGO communicators felt that the caption did not accurately describe the performance and, through a quick chat with an agency representative, had it changed accordingly. Similarly, journalists would sometimes accept NGO staffers' suggestions for what questions they should pose to particular delegations in media briefings. By moving journalists to visibly confront governments on sensitive issues, NGOs extended their politics of public pressure to the press conference rooms, which were otherwise off limits to their campaigners (see 5.2.3). Journalists did not generally object to playing their part in these tactics, maybe because the latter could also mean a good story for them (see also 7.1.2.3).

All three forms of coproduction – (1) information brokerage and provision of final analysis by PR professionals, (2) collaborative composing of photo ops, and (3) continuous informal interactions between both sides – were facilitated by particular parameters of the summit, be it complex subject matter, a shortage of visually appealing images, or the spatial and temporal confinement of the event. These parameters do surely not represent routine interactions between both sides outside such summits. Consequently, HIPS might be conceptualized as exceptional settings facilitating unique constellations between political PR professionals and journalists that see both sides 'coproducing' the event's image distributed worldwide.

7.1.2 The influence of summit coverage on summit proceedings

The following explores the short-term effects that summit coverage has on the summit itself. In general, the role of media in shaping a high-level international political summit, such as the Cancún climate summit, is multifaceted. PR professionals' reflections on how media impact such events can be grouped into three distinct media functions, which can all be made subject to PR efforts: Firstly, as explored theoretically in chapter 2, media presence and attention is a constitutive element of such events. Only through high levels of coverage do they become media events and, hence, occasions of concentrated political and communicative efforts. Secondly, media coverage can set the tone, or wider frame, of a summit. It may determine what is to be expected from the gathering and suggest an overall evaluation of how the talks are progressing. Thirdly, media might function as concrete tools for actors involved in the negotiations. Journalists' coverage – or at least their presence at the site of the summit – can be utilized for putting pressure on or conveying negotiation positions to other actors. I explore PR professionals' reflections for each of these three media functions in the following.

7.1.2.1 Media attention as constitutive element of the summit

Regarding the first function, PR professionals of both governments and NGOs see high levels of media attention as inextricable element of climate summits. They only exist in the eyes of audiences if they are reported on by the media. Thus, varying degrees of media attention might be interpreted by audiences as fluctuations in the importance of these events:

> "If a lot of people don't see that much media coverage on [the summit], then they start to think that it's not that important. For example, comparing to last year, there was not so much media coverage on Cancún in general. And I think that's not a good thing, because then it gets off the radar" (2403, p. 10, 6).

It was pointed out frequently that after the media craze surrounding COP-15 in Copenhagen – a summit that was described as a very exceptional 'mega event' by many –, media interest in the climate change negotiations had dropped drastically and led to much lower levels of coverage on Cancún. Some government communicators seemed even a bit relieved about this, as it allowed for more focused, substantive work and reduced the risk of overburdening the talks with high expectations. NGOs' PR professionals, however, came from the opposite angle and argued that high media attention should be considered a prerequisite for progress in the negotiations. If negotiators on site are under the impression that the whole world is watching, that people everywhere consider their work important, they will show more ambition in finding an agreement. An imagined global audience gives them a sense of mission and urgency. Hence, there are internal and external implications of this first media function: media attention puts the summit on the map for audiences and instigates their interest, which, in turn, provides negotiators with an incentive to act.

7.1.2.2 Media setting the tone of the summit

The first media function is only about the *quantity* of media reporting. It does not say anything about the tone or framing found in coverage but merely refers to the extent to which the event is made visible through the media. In the eyes of NGOs, sparking media interest in a climate summit is something like a precondition for its effectiveness. It becomes a forum likely to produce political progress only by means of prominent media coverage. This is what makes it matter to the world. The *quality* of that coverage, however, is a different question. Media may assess the role and potential outcome of a summit in different ways and may suggest a dominant reading for making sense of the event. The Copenhagen summit, for example, seems to have been framed mainly as a moment of destiny, in which the fate of the world was at stake. As described earlier, for the Cancún summit, most NGOs and some governments attempted to replace that message with a less overburdening frame that also appreciated smaller steps that would not save the whole planet right away (see 6.2.2). Through mechanisms of coproduction explored above (see 7.1.1), this new message found its way into media coverage and seems to have been amplified as the general frame of COP-16.

This, then, describes media's second function in the context of a summit. Media reporting can set a tone, suggest a wider frame, or create an umbrella narrative for the intricacies of the negotiations. This can signal to the world as well as to summit participants themselves what is to be expected of the event. Consequently, the overall frame set by media coverage may become the guideline that negotiators adhere to and against which their work is measured, as one NGO communicator pointed out:

> "The reporting around how no government – or not any serious block of governments – was calling for a full agreement was a self-fulfilling prophecy, of course, or at least contributed to that. It was reaffirming the fact that everyone was looking for these sorts of building blocks – and that is what we got! And it allowed for that to be an achievement rather than a failure. Because you can look at it in two ways: If you are calling for a full agreement and you only get building blocks, it is a failure. But if you are calling for building blocks and you get building blocks, it is an achievement – even though it's the same outcome! So that reporting, even ahead of time, but also during the negotiations, I think contributed to that actually happening" (9557, p. 8, 24).

From an NGO perspective, the strategy of utilizing the second media function for setting a more realistic tone seems to have resembled something close to a dilemma: calling for steps that were actually regarded as insufficient, while knowing that calling for sufficient steps was unrealistic. Out of these two evils, all NGOs in the sample, except for Friends of the Earth, chose the first and promoted the 'building block' or 'step-by-step' frame. This, then, lowered the bar for negotiators and made a positive assessment of the summit outcome more fitting.

Besides this particular frame, another broad message seemed to dominate summit coverage in the eyes of PR professionals, both of governments and of NGOs. This message was not about the negotiation substance itself, but about how the negotiations were led. As described above, the Mexican government had given out 'transparency' as the keyword of its COP presidency (see 6.2.1), reiterating whenever possible that under its leadership, the talks were open and inclusive and that no secret negotiations behind hidden doors were taking place. This became a meta-message, or meta-frame, of summit coverage, as one NGO communicator illustrated:

> "The Mexican government did a really impressive job of really emphasizing the process that they were undertaking as an inclusive, transparent et cetera process. And that was reported on, where I think media were saying that the negotiations were being conducted in that way […]. I think that the messaging that the Mexican government did around that was pretty impressive from a communications perspective. Every single time they spoke, they were talking about it. And it was pretty solid message discipline, if you want to call it that" (9557, p. 7, 33).

According to that particular PR professional, the omnipresence of the transparency theme in summit coverage and its embracing by negotiators built the foundation for the ad-hoc drafting of the Cancún Agreements in the final hours of the summit. Only because parties were ensured over and over again through interpersonal communication and media coverage that they were all on the same page, that there were no hidden pacts, and that no one was invited for special negotiation rounds, the Mexican presidency could quickly unite nearly all parties (except for Bolivia) behind the communiqué.

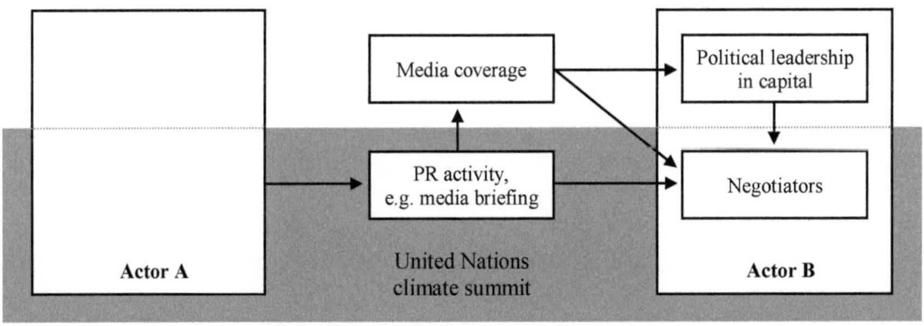

Figure 7.1: Different avenues for media involvement in the communication between two summit actors

Hence, in this case, media – carrying out their second function – did not provide for a particular expectation towards the summit outcome but, on a meta-level, created an atmosphere, or breeding ground, conducive to finding agreement. One NGO communicator stressed the importance of such 'process stories':

> "[The media] can not only look at the negotiations themselves in terms of content, but also 'are things going as they are supposed to go?', 'do non-governmental organizations get enough room to be heard?', for example. They are also an outside check to see if everything goes the way it should go at negotiations like that" (2403, p. 10, 15).

By establishing expectations regarding the substance and process of HIPS, media coverage may impact not only how summit outcomes are ultimately evaluated, but also in which atmosphere they are attained.

7.1.2.3 Media as negotiation tools

The third function of media in the summit context refers to their exploitation as instruments of negotiation by political actors. PR professionals of both actor types repeatedly pointed out throughout the interviews that at climate summits, media are widely acknowledged as channels of negotiation and are commonly used by parties and NGOs. Different strategies of engaging journalists for negotiation purposes could be detected, which could be grouped into three distinct avenues of media involvement (see figure 7.1):

Firstly, if an actor deems its communication with another actor – be it in direct exchange or at the negotiation table – to be ineffective, it has become popular tactic (especially by NGOs) to involve journalists. The motive here is not so much to instigate media coverage about the conflict right away, but to use the deliberate informing of journalists – who *potentially* could make public whatever they know – as a first means of exerting pressure. Demands vis-à-vis an actor could then simply be voiced in the presence of journalists; however, the latter might also get lobbied specifically to incorporate them in their own inquiry with the actor. Under the label of 'planting questions', this is at times done by NGOs at governments' media briefings. Due to a restriction of access to the press conference rooms (see 5.2.3), NGO representatives wanting to confront a government over a certain issue in the presence of journalists cannot do so on their own but have to find journalists who raise the issue in the briefing. At the same time, for the targeted government, a demand built into the question of a journalist, who is always in search for stories, carries more significance than if it comes along as the usual NGO complaint. Here, journalists are used not so much as producers of media coverage, but rather as authoritative witnesses and intimidating messengers, as NGO communicators explained:

> "If someone is being bad, you make sure you plant questions into those press conferences [...]. Sometimes it is the coverage, but sometimes it's more just the questions [journalists] ask" (0417, p. 8, 9).

> "It's giving journalists the right questions to ask, giving them the right story ideas, telling them that that's burning, these guys are trying to do something nasty. So that's direct, that's not so much the reporting back" (1634, p. 6, 10).

In the two other avenues of media involvement, media coverage is sparked purposefully in order to put pressure on or convey a position to another actor. For some parties, this has become common practice, as one NGO PR professional observed:

> "Many of them engage in something that I refer to as 'PR diplomacy'. I think what's happening in the press conference rooms here is often – not always, but often – as important as what's happening in the negotiation rooms, because it seems like some countries at least like to negotiate via press conferences" (9841, p. 8, 10).

In this 'PR diplomacy', the role of media might on one hand be viewed as that of a simple intermediate channel allowing transfer of messages from actor A to actor B (Carpenter, 2001, p. 319; see also Gilboa's [2001] 'media diplomacy'). On the other hand, media act as amplifier, making messages accessible to a much wider audience and thereby allowing for such strategies as public shaming or isolating, as highlighted by one NGO communicator:

> "You can really influence the politics, the negotiations with sharp analysis in the media, with good quotes. People really do look at that, and they are afraid of bad publicity" (2403, p. 10, 10).

Regardless of the exact motive, the instigation of media coverage at a summit is usually meant to affect either the negotiators right on site (the second avenue) or, through a two-step process, their political leadership back in their capitals, which might adjust negotiators' mandate in response to what is reported on the summit. (This third avenue could be seen as a kind of 'reversed boomerang effect' – see 3.1.3.). When political actors target journalists in their communication activities, it is usually to reach one of these two groups, as two NGO communication professionals outlined:

> "You have a little circular thing that goes on with the political/environmental journalists that cover the meeting: they then influence, hopefully, through their media back at home, the delegation. And that's it

really, because the delegation is also monitoring what their journalists at the meeting are saying – and so are their bosses back home! […] Getting the national coverage back home is really important" (0417, p. 9, 16).

"Public communication efforts can force governments to define new parameters for their negotiators from their capitals" (9557, p. 5, 30).

Instigating media coverage for the purpose of making negotiators and/or political leaders subject to public pressure (or at least for hinting at the possibility of public pressure arising) is a strategy mainly carried out by NGOs. However, governments also use the media actively as tools for shaping negotiations. Especially the exchange of negotiation positions or the testing of initiatives is often done through summit coverage, as one NGO communicator explained:

"So if one party says something in a press conference, then that could be a signal that they are ready to move and they are doing it as a test balloon and they hope that two hours later, when the other party is having their press conference, they might respond and send their message out somehow. And then they know 'okay, there was a response; they are also ready to move, now let's bring it to the negotiating table and see'. That's often how it works in my view" (9841, p. 8, 32).

In turn, negotiators' statements in the media have become an import source for political intelligence-gathering. Parties and NGOs closely monitor summit coverage for what could be deciphered as clues to how a party's position is developing, as both state and non-state communicators acknowledged:

"You monitor what other people are saying in the media and other negotiators putting across certain views that begin to give you a sense of how they would approach certain issues. So I think in a big way, it helps in terms of showing how, what approach the negotiators from different countries are taking into those particular negotiations" (5095, p. 6, 7).

"Press conferences can be full of important signals, whether that's a move forward, a move backward etc., and nuances count a lot in these negotiations, because they are in such a bad shape that a lot is about atmospherics, about vibe, about whether they behave and engage in good diplomacy or whether they don't. That's why nuances are so important" (9841, p. 8, 21).

In certain contexts, talking to the media might also have counterproductive effects, however. A government communicator recalled a situation from the Copenhagen summit, where the exchange of negotiation positions via the media did not facilitate but hinder agreement:

> "What happened is that parties or blocks went to the media before, and when their position was out, others that probably would have gotten into that same agreement didn't go for it. They thought that they were not included" (8990, p. 3, 30).

It becomes apparent that in one way or the other, media can play an important role in shaping the political process of a climate summit. Their strong presence at the event makes it easy for governments and NGOs to view them as potential tools and to integrate them in their strategies. While this third media function is surely not limited to summits – the running of political negotiations through the media might also take place in a context of every-day political communication at the national level –, it does have a fast-paced, self-reinforcing side to it at HIPS.

7.2 Short-term discursive opportunity structures

The dynamic, action-filled nature of HIPS, where negotiation positions or actor constellations may change overnight, forces PR players to stay flexible and be prepared to quickly modify their communication in response to current happenings. These developments might come in form of short-term discursive opportunity structures (see 3.3.2 for the theory behind the concept) that have to be taken into account by actors in their summit communication. In short, summit dynamics might provide political actors with unforeseen themes, channels, or targets of communication that they had not foreseen in prior planning but require on-site responses.

The emergence of one particular short-term discursive opportunity structure and its subsequent integration in actors' communication strategies are traced in the following. The example vividly illustrates the rapid interplay of different actors at the Cancún summit and how one actor's move can bring about others' responses, which might again affect the initiating actor (though, in this example, only indirectly). It emphasizes the fact that ex-ante planning of summit communication is always contingent on the actual unfolding of the event. By means of this example, the occurrence of short-term discursive opportunity structures at HIPS can be confirmed as constituting a significant factor shaping communication strategies – as hypothesized in my conceptual model.

The start of developments was a statement by Japan's chief negotiator, Jun Arima, on the second day of the summit (Daiwa Foundation, 2011; Vidal, 2010a). In what was widely perceived as usually harsh rhetoric, the official clearly rejected the idea of extending the binding limitations on greenhouse gas emissions inscribed in the Kyoto Protocol:

> "Japan will not inscribe its target under the Kyoto protocol on any conditions or under any circumstances" (Vidal, 2010a).

Up to that point, provisionally extending the Protocol until a new agreement would come into effect had been considered a viable, if not vital, option by some parties, especially in the developing world (Black, 2010; The Japan Times, 2010). While Japan's skepticism towards prolonging the Protocol was already known before the Cancún summit, the rigor and determination expressed in Japan's language was still met with astonishment:

> "The move came out of the blue for other delegations at the conference. 'For Japan to come out with a statement like that at the beginning of the talks is significant,' said one British official. 'The forthrightness of the statement took people by surprise' (Vidal, 2010a).

> "Japan's statements […] jarred the negotiations, a realm where language without caveats […] is a rarity" (Revkin, 2010).

Japan's move – and choice of language – was also met with surprise by NGO communicators:

> "I don't think we'd really expected Japan to actually walk away from Kyoto so publically in the middle of the meeting. I don't think that was really expected" (0417, p. 4, 27).

> "Japan was talking about something they'd been talking about a million times before. It was just so significant that they said it in harsh words […] on opening day, when usually nobody wants to poison the atmosphere" (9841, p. 8, 27).

> "While they had made that decision of not wanting to be part of a second commitment period [of the Kyoto Protocol] relatively clear ahead of the negotiations, their strategy and posture, which was quite, I'd say, aggressive from day one of the negotiations, was something that we had to react to pretty quickly" (9557, p. 2, 9).

Japan's announcement triggered strong reactions from other parties:

> "Nations looking for a new deal have launched a diplomatic assault on Japan in the hope of softening its resistance to the Kyoto Protocol. As many as 20 world leaders are in line to phone Prime Minister Naoto Kan to ask for a change of stance. Japan's position is seen as the single biggest barrier to reaching a deal" (Black, 2010).

> "A senior Japanese negotiator today said that it had come under intense diplomatic pressure to soften its stance at the UN climate talks in Cancún and admitted that it was causing a 'big problem' for the negotiations. But he repeated the country's position that it would not compromise on its refusal to sign up to a second commitment period of the Kyoto protocol – the international treaty that legally binds rich countries to cut emissions" (Vidal, 2010b).

Reacting to Japan's announcement became the predominant mission of NGOs' PR professionals. While Japan had not been on NGO communicators' radar as a particular target or theme of communication prior to COP-16, it had now recommended itself as focal point. The larger strategy quickly conceptualized by NGOs for responding to Japan's move could be summarized as 'pointing out the bad guy'. Usually, such an approach would involve intensive press work in the targeted country in order to convey to the domestic public the impression that its governments was pursuing isolationist policies and to spark domestic opposition to that. However, in the case of Japan, NGOs refrained from choosing that strategy, as they were under the general impression that due to general skepticism towards civil society actors in that country, their Japanese chapters did not enjoy a particularly high credibility domestically. Experience had shown, for instance, that press releases by Japanese NGOs were often ignored by domestic media. Hence, working through national offices in Japan to mount *domestic* pressure on the Japanese government was deemed ineffective. NGO communicators in Cancún instead exploited the country's assumed susceptibility to *international* demands. In contrast to other countries (such as the United States), Japan was regarded as being highly attentive to what the world thinks about it:

> "This theory of [...] being shamed on the international platform is actually one that does have an effect [in the case of Japan]. And so getting international voices to really criticize heavily publicly a position that the Japanese government is making or is having in a negotiation can actually affect the outcome of their position" (9557, p. 6, 19).

> "If it doesn't come from Japanese but from foreign voices, it might work! [...] You need to know your ways and channels if you want to reach a specific market, a specific country!" (9841, p. 17, 23)

Consequently, NGOs' response strategy to Japan's move had an explicitly international orientation. The idea, essentially, was to shame the country on the international stage, to make clear that the country had become the outlaw in these global talks – a subject to everybody's annoyance and ridicule:

> "We are making fun of Japan! And we just want to tell them that they can't get away with it, and we want to make them look stupid" (1625, p. 7, 15).

> "So it was clear who the good guys are and who the bad guys are! And that was the intention, because Japan of course tried to rationally explain why they were taking this position – and if you look at it rationally, they might even have some good arguments. But behind their rational arguments, there are different political agendas, and it was important to point out that they are not just acting on the basis of a reasonable argument and that they are indeed to be criticized for the position they take!" (9876, p. 6, 26)

This overall strategy was turned into several concrete activities. Behind the scenes, NGOs were requesting their national chapters to ask politicians in their capitals to call the Japanese government and to underscore its isolated position. This effort certainly contributed to the 'intense diplomatic pressure' that Japan was exposed to (see above). At Cancún, NGOs were also devoting their media relations to drawing journalists' attention to Japan's behavior. The communicator in charge at WWF International recalled that a colleague from WWF Japan, together with a press officer, "gave interviews for almost 40 minutes [...], partly for groups of journalists, like huddles" (1634, p. 1, 23) when the Japan story broke. Friends of the Earth International instantly published a press release urgently calling upon Japan to "reconsider its position and stop stalling climate talks which have hardly even started":

> "Japan's move to drop out of the Kyoto treaty shows a severe lack of recognition of its own historical and moral responsibility. With this position, Japan isolates itself from the rest of the world. Even worse, this step undermines the ongoing talks and is a serious threat to the progress needed here in Cancun" (Friends of the Earth International, 2010, ¶ 3).

Several NGOs devoted at least one of their regular media briefings to confronting Japan on its move. GCCA also attempted to draw attention to Japan by adjusting its social media work accordingly and informing fans and followers via Facebook, Twitter etc.

In addition, NGOs conducted a series of multifaceted campaigning actions targeting Japan, which vividly illustrate the wide repertoire of communication activities that NGOs may resort to. First of all, Japan was awarded the Fossil of the Day award (see table 6.1) on November 30 (as the only awardee on that day), accompanied by a press release blaming the country for "aggressively denying the future of Kyoto in opening plenary" (CAN International, 2010a, ¶ 2). On the day after, the summit newsletter *ECO* featured a fake contact ad, mocking the Japanese delegation as "just looking for a good time in the Cancun sun" (Eco,

2010) and thereby setting the issue on the agenda for the summit grapevine (see figure 7.2).

Cancun Personals

ANNEX 1 COUNTRY SEEKING
TREATY FOR NO-STRINGS ATTACHED
HOLIDAY ROMANCE IN MEXICO.

Currently struggling with a 13-year relationship, just looking for a good time in the Cancun sun.
Likes: excellent food, movies, comic books, robots and big industry.
Dislikes: commitment, cooperation, compliance, science and targets.

If interested please email:
scared_of_commitment@awg.kp

Figure 7.2: Fake contact ad in December 1, 2010 issue of the *ECO* newsletter (Eco, 2010)

A day later, on December 2, 2010, GCCA organized a smaller photo op at *Cancunmesse*. While the action was directed primarily at conference participants (who had to pass by the scene on their way from the security checks to the shuttle buses towards Moon Palace), it was also recorded by news agency photographers and their images used by media outlets around the world (see figure 7.3). The stunt featured young campaigners in 'I ♥ KP' ('I love the Kyoto Protocol') T-shirts holding a red, heart-shaped frame of oversized dimensions, mimicking the backdrop of kitschy wedding pictures. Conference participants could then have their picture taken with the frame and thereby express their love to the Protocol (symbolized also by the cut-out letters 'KP' that participants could show to the camera during the picture-taking). Additionally, sounds from the Beatles song 'All You Need Is Love' filled the scenery. The stunt is a fitting example for NGOs' common tactic of integrating serious motives (such as sustaining pressure on Japan) with surprise and humor.

Figure 7.3: 'I ♥ KP' photo op by GCCA on Dec. 2, 2010 (author's collection)

The most noteworthy campaigning action occurred on December 10, 2010, the final day of the summit, when GCCA/TckTckTck, in cooperation with the online activist network Avaaz, published an advertisement in the international edition of *The Financial Times* (and in the summit newsletter *ECO*), which was to highlight the ill-fated course pursued by the Japanese government. The ad, which roughly occupied one sixth of a newspaper page, imitated the poster of the 2001 Japanese animated film 'Spirited Away', "the highest-grossing film in Japanese history" (Wikipedia, n.d., ¶ 3). Designed in similar style as the movie poster (see figure 7.4), the ad showed Japanese prime minister Naoto Kan being confronted by a huge tidal wave. As a clear reference to the real-life movie title, the headline above this motif asked 'Washed Away?'. The complete text of the ad read as follows:

> JAPAN presents a THREATENING TO ABANDON KYOTO film:
> CLIMATE TREATY: WASHED AWAY?
> Japanese Prime Minister Naoto Kan is living in a fantasy-imagining he can refuse a new Kyoto Protocol commitment period without wrecking hopes for a global climate treaty. As UN talks in Mexico bog down, the world needs Kan to wake up: if he abandons Kyoto, the climate treaty will be washed away!
> AVAAZ & TCKTCKTCK.org, in association with ALL LIFE ON EARTH, at this WORLD-IN-THE-BALANCE moment, urges NAOTO KAN and

> THE GOVERNMENT OF JAPAN to recommit to the KYOTO PROTOCOL to prevent us all from BEING WASHED AWAY."
> (TckTckTck, 2010)

In a news article on the TckTckTck website, GCCA added some meaning to the imitation of the 'Spirited Away' poster:

> "In 'Spirited Away', a little girl follows her parents down a dirt road ending up in a fantasy world. 'Like the girl [...] Japan is right now at cross roads', said [Paul] Horsman [= GCCA's Campaign Director]. 'One path leads to fantasy land, where the world's hopes for a binding treaty are washed away. The other leads to true leadership and a safe climate future'" (TckTckTck, 2010, ¶ 6).

In other interpretations, the ad might also be seen as a reference to the 2000 Hollywood movie "Cast Away" starring Tom Hanks as the survivor of a plane crash stranded on an isolated island and struggling to make his way back into civilization – a plot that might have reflected the course of the Japanese government in the eyes of NGOs.[12] Finally, the large tidal wave depicted in the ad could also allude to the fact that Japan, as a group of islands surrounded by oceans, is exposed to the forces of nature to a greater extent than other countries and should hence have a clear interest in tackling climate change.

As in the wedding-picture set/Beatles song photo op described above, the ad exploits references to principal products of pop culture – here: movies – for sparking attention and conveying surprise. This very tactic had already been carried out at the 2007 UN climate summit (COP-13) in Bali, where CAN International focused its efforts on isolating the US for its position in the negotiations of the Bali Road Map. Part of that strategy was to move the two main allies of the US in the negotiations, Japan and Canada, to a point "where they weren't comfortable backing up what the US was doing" (3148, p. 4, 17) and would hence turn their backs on the US:

> "We did a major communications effort from Bali, but also in the countries – Japan and Canada – to really shame them by associating them with President Bush at the time and to really encourage them to listen to the public that was calling for them to step back from the position that they were supporting the Bush administration: [...] One of our member organizations took out an ad in a Japanese newspaper that was designed to look like a 'Titanic' movie poster with three heads – the prime minister of Japan, Canada, and the president of the US – to associate Japan with that tri-factor, that was obviously going to be shameful. And we saw them step back; we saw them

12 I am indebted to my colleague Anja Peltzer for alerting me to this pop cultural reference.

> change their positions or at least not vocalize them in a corruptive way" (3148, p. 4, 17).

The member organization behind the publishing of the 'Titanic' ad (see figure 7.5) was once again Avaaz; the paper carrying the ad, however, was not a Japanese publication (as posited in the quote above) but appeared to be English-

Figure 7.4: Left: protest ad in December 10, 2010 issue of the *Financial Times*' international edition (TckTckTck, 2010); right: movie poster for the 2001 Japanese animated film 'Spirited Away' (Wikipedia, n.d.)

language *Jakarta Post*, one of the most prestigious newspapers in the host country of Indonesia (Canwest News Service, 2007). Avaaz did indeed pride itself for having changed Japan's position in the Bali negotiations through the release of the ad. On its Flickr profile, the organization presented an article by Japanese newspaper *Asahi Shimbun* published on January 6, 2008 (roughly three weeks after the end of the Bali summit), which – as the NGO claims – describes the direct impact the 'Titanic' ad had with Japanese government leaders. Avaaz offers the following translation of the article:

> "Japan went to Bali with an unclear position, and came under severe criticism from NGOs. 'Japan, the mother of the Kyoto Protocol, is trying to throw the protocol away', environmental NGOs reacted sensitively. […] However, in the aftermath of Bali, things appear to be shifting as a result of what is being described as 'the Bali Shock', the fact that the government was not counting on such rigorous criticism from NGOs:
>
> On the 27th, there was a global warming tactics meeting for four senior cabinet ministers. […] In this preparatory meeting, the Environment Minister Kamoshita pulled up an Indonesian newspaper, the *Jakarta Post*. Faces of Prime Minister Fukuda, President Bush and Prime Minister Harper appeared, covering the entire back page of a special section on the Bali negotiations: […] 'No targets, no icebergs, just global disaster – coming soon. World don't give in!' It was an ad of an environmental NGO.
>
> Minister Kamoshita reacted: 'Japan is regarded as resistance power by the world. Can we get away with this?'[13] 'Why don't we propose an emission target?' asked the Chief Cabinet Secretary, a comment that was seconded by the Environment Minister and the Foreign Minister. […] This talk convinced the Prime Minister to develop a national mid-term goal." (Flickr, n.d.)

Having successfully caused the 'Bali shock' for the Japanese government, NGOs in Cancún seemed to strive for the same effect once more: to pillory the Japanese government by means of an ad in an internationally visible outlet, to brand it as set on a course of isolationism and disaster, and, thus, to eventually change its position at the negotiations. In order to facilitate the impact of the 'Washed Away?' ad, GCCA even had a thousand copies of the respective *Financial Times* issue distributed at Moon Palace and *Cancunmesse* (TckTckTck, 2010). Hence, the organization's idea for exploiting the short-term discursive opportunity structure created by Japan appeared to include a resorting to communication strategies and activities that had worked effectively at the Bali summit.

How did Japan respond to the headwinds it faced at Cancún summit? As mentioned by several NGO communicators in my interviews and publicly

13 In other Internet sources, this statement is translated even more to the point: 'The world sees Japan as a force resisting change! Are we okay with this?' (Indymedia, 2008, ¶ 3; News Around Us, 2008, ¶ 8)

acknowledged by Japanese government officials, the country came under concerted pressure. In addition to NGOs, especially representatives of the G77, the group of developing countries, condemned the country for its move and stressed the need for an extension of the Kyoto Protocol, which, under the so-called Clean Development Mechanism (CDM), provides them with special funds for sustainable development projects (Vidal, 2010a; Black, 2010). Japan appeared to be very attentive to all these complaints voiced by parties and NGOs; according to one NGO communicator, the 'Washed Away?' ad in the *Financial Times* even made its way onto Japan's cabinet table.

Figure 7.5: Protest ad in December 14, 2007 issue of the *Jakarta Post* (Center for Educational Design and Communication, 2007)

Japanese reactions, however, turned out to be of a different nature than hoped for by protesting actors. Instead of *changing* Japan's position, the diplomatic pressure by parties and the isolation campaign of NGOs seemed to even *reinforce* it. In the days after the initial announcement, which immediately triggered a wave of protest, Japanese officials kept on repeating the country's position – sometimes put in even harsher words or accompanied by explicit remarks that protest and pressure will not bring about to a change in positions. Even two days before the end of the summit, when all communication activities targeting Japan (except for the 'Washed Away?' ad) had already taken place, the ambassador for civil society in Japan's foreign ministry, Akira Yamada, underscored his country's iron-hard position:

> "We are not moving. This is a fact. Many people have had the illusion that Japan might change its position. Well, we are sorry, but we are not going to. There is 0% possibility" (Vidal, 2010b, ¶ 3).

It seemed like Japan had prescribed itself absolute immunity from any external pressure, as if the government had expected nothing else but such reactions and had therefore aligned all his spokespeople to staying firm and unshakable ahead of the summit. Evidence for such a conscious 'hardening' of the country's position might be found in an article published by the *Japan Times* on November 27, 2010 – two days before the start of the Cancún summit. In the article, the deputy minister for global environmental affairs at the Japanese environment actually predicted his country to be made subject to isolation campaigns at COP-16 but expressed the determination to uphold the much-criticized position nonetheless:

> "Even if the Kyoto Protocol's extension becomes a major item on the agenda at Cancun and Japan finds itself isolated over it, Japan will not agree to it" (The Japan Times, 2010, ¶ 3).

The Japanese government had obviously foreseen the turmoil that its announcement would create at Cancún and had hence prepared itself for a rough ride. This means that while NGOs' ad-hoc communication strategies did not significantly influence Japan *during* the summit, their anticipation by the Japanese government *prior* to the event seemed to have an effect: It contributed to the development of an unmistakably clear message and the strict adherence to message discipline by all delegation spokespeople. It appears that prior to the Cancún summit, the Japanese government had dug out its lessons from Bali and – in an effort to avoid a 'Cancún shock' – had deepened its trenches and heightened its walls. This might be seen as a long-term learning effect by the Japanese government on how to deal with NGO opposition.

After the summit, one NGO communicator wondered whether Japan's bold statement could have been predicted, whether they had missed out on signals hinting towards the use of such strong rhetoric:

> "Should we have known that Japan was going to come out with such a strong position? And could we have done something ahead of time to change that? I think that that's a question that we have been thinking about. And what the timing of our communications are and how we target them, are lessons that I think we needed to take forward" (9557, p. 5, 14).

Other NGO PR professionals concurred that Japan's move came unexpected indeed and constituted an important short-term factor influencing strategy choice at the summit:

> "I don't know that we necessarily had a plan to go after Japan, but it depends on what is actually happening inside" (2419, p. 8, 5).

> "I guess the Japanese bit was the biggest surprise. […] We were very negative, I think, certainly much more negative than our framing would have allowed us – we went off frame! […] We couldn't really do anything else at that point" (0417, p. 4, 33).

As illustrated by these quotes, the emergence of a short-term discursive opportunity structure at a summit can cause actors to (partially) abandon their strategies (in this case, the explicitly positive and appreciative messaging around the negotiations) and adjust their communication to new realities of summit dynamics (here, by resorting to public shaming and other forms negative campaigning). This shows that not all factors shaping communication strategy at HIPS are inscribed in the actors' communication structure or subject to long-term planning. The unique way in which a summit unfolds constitutes equally significant short-term factors.

7.3 Political actors' learning effects

Having explored the first feedback cycle of my conceptual model – the mechanisms through which political actors' communication strategies may set off short-term effects at the summit that again affect their strategies –, I now explore actors' long-term learning effects: How do governments and NGOs assess their strategies at the summit post hoc? Which lessons do they take home? This second feedback cycle focuses our attention on actors' readiness to evaluate their summit communication and deduce strategic insights for coming events. Up to this point, we have come across examples of Japan learning from its COP-13 experience and hence immunizing itself against public pressure at COP-16 and of major NGOs learning from their COP-15 experience and hence rolling out a more upbeat messaging strategy at COP-16. Provided that actors make available sufficient reflective resources, we can hence assume that COP-16 might also have brought about lessons to be considered when developing strategies for summit communication in the future.

Overall, NGOs seemed to be more ready than governments to evaluate and reflect upon what and how they did at the Cancún summit. This should also be seen in the context of NGOs appearing to possess a clearer understanding of what they want to achieve communicatively, sometimes even in the form of concrete quantitative and qualitative benchmarks. Hence, if an organization connected the event with particular communication objectives, it appeared more likely to check up on their fulfillment. Conversely, delegations lacking precise ambitions in their summit communication were more superficial in evaluating.

Governments' learning effects

Out of the governments in my sample, only *Brazil* seemed to have put in place a more thorough process for evaluating its summit communication. All instances of media coverage on Brazilian activities or the Brazilian position were documented, and the tone of coverage was analyzed in a report concluding that it was "positive and consistent, offering key insights, analyses, and commentary from our spokespersons on site in top international news wires, print, broadcast, and online news, and trade outlets" (6775, p. 1, 8). The Brazilian government representative voiced his satisfaction with how their message resonated with the media and saw it as an important step in 'greening' the nation's image. The overall evaluation was hence rather positive. The only aspect that was viewed critically concerned the timing of the daily media briefing: Its starting time at 6 pm gave journalists working for morning newspapers and evening newscasts in Brazil and Europe little time until editorial deadlines. In response to this lesson, time zone differences might be considered more strongly when scheduling the daily media briefing at future summits.

The communicators of *India* and *South Africa* also expressed positive appraisals, although that did not always seem to be backed up by systematic evidence. The Indian government representative cited headlines like 'Jairam Ramesh praised for building bridges at Cancún' or 'India enjoys place in the sun' as indicators of worldwide appreciation by governments of the role that India – and particularly its environment minister – played at COP-16. The South African communicator pointed to strong media interest in the readiness of South Africa to host the next climate summit and described that coverage as rather positive – especially since the country had already demonstrated its capability of organizing such events by hosting the 2010 FIFA World Cup. For him, two observations from Cancún were considered important lessons for the hosting of the Durban summit: Firstly, a prior assessment of what can realistically be achieved and an appropriate management of expectations are key to making the summit a success. Secondly, hosting a summit should also be seen as an opportunity to make public to the world one's own climate actions and also to market the country as a tourist destination – just as the Mexican government did in Cancún.

As current host government, the *Mexican* government appeared to be rather enthusiastic about what it saw as very positive media feedback to the summit. While it had explicitly downplayed the likelihood of 'something big' coming out of Cancún beforehand and had instead foreseen some conflict and dissatisfaction among parties, it described the outcome and the coverage about it as "better than expected" (0240, p. 1, 3). Summit coverage was considered a "boost of good news about the country, about what Mexico is beyond these issues of violence and narcotics" (0240, p. 1, 3) and hence a "refreshment" (0240, p. 3, 10) of the country's international image. Concrete lessons were drawn primarily regarding the technicalities of hosting large events. Especially the splitting up of and long transfer times between venues were met with internal criticism and mentioned as one aspect to be reconsidered in the future (see 5.2.2).

Except for minor points regarding technicalities, such as the furnishing of information booths, little could be found out about internal evaluative or reflective processes by the two industrialized countries in the sample, *Germany* and the *US*. As established players in the UNFCCC process and on the word stage, it seemed like their summit communication were following more traditional routines and less part of an explicit agenda to take a stand or to increase visibility. In other words, they appeared to run their summit communication like they have always run them – without ambitions to scrutinize them for room for improvement and to initiate major changes. Indeed, one journalist described them as "old hands at doing this" (6668, p. 4, 27), and one NGO communicator recalled that their communication at climate summits in the past "were done the way they do them now" (9841, p. 12, 23). Hence, the long-term feedback cycle constituted by internal evaluation and subsequent adjustment of strategy should not be considered very pronounced for these two actors.

NGOs' learning effects

In contrast to governments, non-governmental organizations tended to spend more time evaluating and reflecting upon the effectiveness of their strategies.

Especially for the two NGO alliances, *CAN International* and *GCCA*, which in Cancún oversaw the roll-out of a new messaging strategy that abstained from making overly ambitious and unrealistic demands and instead focused on appreciating small steps and highlighting climate action outside the UNFCCC process, assessing the success of that new approach was important. Representatives from both organizations voiced satisfaction with the extent to which that new frame had found its way into summit coverage. In their perception, media mainly went for the kind of 'balanced' assessment of the summit outcome – not too gloomy, but also not overly positive – that they had facilitated. It confirmed to them the importance of expectation management as part of a summit communication strategy. On a meta-level, as the CAN representative pointed out, the shift in strategy from the Copenhagen summit might also illustrate NGOs' capacity to stay flexible and adapt, which is a success in itself. The GCCA communicator, however, voiced his regret at the rift in the NGO community that became visible between mainstream and more 'radical' groups and hoped for more unity in Durban.

Greenpeace International, as one of the largest contributors to the new messaging strategy, expressed some smaller doubts about the new approach. While being fully committed to the new frames of 'choice' and 'opportunity', on site in Cancún, the Greenpeace communicator experienced some political journalists' hesitation to incorporate these new perspectives into their coverage. While they acknowledged the new tenor coming from NGOs, some of them found it too abstract and carrying too little substance. In general, according to that particular PR professional, the playing around with new messaging strategies has just started, further work will have to be done in tailoring it to journalists, but also in promoting it internally.

The optimistic messaging strategy carried out by the alliances and Greenpeace International stood in clear contrast to *Friends of the Earth*'s stance on the summit. In cooperation with the *Via Campesina* movement and the Bolivian government, the NGO clearly denounced the summit communiqué but, in its communicators' impression, was unique in choosing that kind of rhetoric. While the benchmarks for its media relations activities were all met and media response was perceived as high in general, the tone of coverage was not at all in line with what Friends of the Earth had hoped for. The NGO had arrived in Cancún with rather low expectations but a clear mission to prevent positive spin of what it considered 'false solutions'. When it saw exactly that happening – and fellow NGOs even contributing –, it caused some disappointment and frustration within Friends of the Earth and even triggered some doubts as to whether being present at COPs with dedicated communication staff was still an efficient use of resources. As FOE's positions were not given much consideration in Cancún, so was the reasoning, media relations could also have been run from the international secretariat and national offices at much lower costs. This, however, was just a thought expressed in the Cancún aftermath. FOE's position, in the eyes of its communicators, had certainly reinforced the organization's image as a more 'radical' and progressive actor within the environmental community.

While *WWF International* did not appear to integrate the new messaging strategy into its summit communication as much as other actors did, it still tested a new approach at COP-16, focusing almost entirely on working closely with journalists from 'high-quality' outlets and providing them with background and analysis. In the impression of WWF International's communicator in charge, this strategy of "educating journalists" (6282, p. 1, 10) worked well, even though he was not able to assess the influence it had on the quality of summit coverage. What had been measured in hard numbers, however, was the number of articles triggered by NGOs. Here, WWF came in third, behind Greenpeace (which had more than twice as many) and Oxfam (which was only marginally ahead). According to the WWF representative, this was unusual for the communicator, as its organization would normally be "back-to-back with Greenpeace and better than Oxfam" (6297, p. 1, 12). He saw the reason for this shift in shares in the nearly complete absence of photo ops conducted by WWF International. This did not only deny media reasons to report on the NGO but also had indirect effects: Being less visible due to a lack of stunts decreases chances of being requested as provider of quotes and expertise. Apparently, NGOs need to be noticeable and maintain 'brand visibility' on site in order to recommend themselves as possible sources to journalists. At the time of my concluding interview, it appeared to be a somewhat open question for WWF which path to follow in the future: that of quieter behind-the-scenes media relations, aimed at shaping the *quality* of coverage, or that of being loud and visible, geared towards triggering a maximum *quantity* of coverage. This also depends on budgetary developments in the future, as the staging of photo ops requires additional material and personnel resources, which are not needed for the 'educating journalists' strategy.

Further operational lessons for NGOs

NGOs did not only evaluate their broader strategies but also scrutinized their structures and day-to-day processes for lessons to be taken into account at future summits. In terms of communication structures, the GCCA communicator found it remarkable that dedicated communication capacity was almost entirely limited to large NGOs, while smaller organizations did not bring along own communicators (even though some of them arrived with ambitious plans for publicizing statements, launching publications etc.). GCCA then had to provide ad-hoc PR support to these actors but became a bit overstretched in the process. As a lesson, the alliance might work "towards having more capacity on the ground to support those smaller partners" in the future (2176, p. 11, 18). Friends of the Earth International also acknowledged some obstacles in its communication structure (such as limited involvement of policy experts or long chains of approval) that complicated the preparation of press releases and planning of media briefings during the summit. In response, ideas for streamlining these structures were formulated.

In terms of processes and concrete instruments, GCCA and Greenpeace International both voiced their satisfaction with the media impact of their photo ops, especially of their joint operation involving an oversized life ring on the last day of the summit. As a consequence, Greenpeace saw its reputation as best picture provider restored (which had apparently suffered in Copenhagen, where the NGO could not do many stunts). Greenpeace also pointed out the good experience made in Cancún with providing journalists with a rapid-response analysis of the final communiqué in the final hours of the summit. Due to the appreciation and exploitation by journalists, this communication activity was regarded a top candidate for repetition at future events. It certainly contributed to Greenpeace's mentioning in seven percent of all summit coverage (according to the NGO), more than twice as much as any other international NGO.

7.4 Chapter summary

This chapter dealt with the dynamic, interactive side of public diplomacy at HIPS. I first investigated the effects of summit PR on the work of journalists at the summit and detected a constellation of coproduction involving both sides. Through various low-key and informal interactions on the site of the summit, PR professionals and media representatives collaboratively shape the public image of the event as conveyed by summit coverage. This is visible in, e.g., PR professionals' information brokerage or journalists' contributions to the crafting of NGOs' photo ops. PR professionals generally conceive of summit coverage as a powerful influence on summit proceedings. It sets expectations – and hence the standards by which summit outcomes are evaluated – and may serve as concrete tool in negotiations, e.g., for creating public pressure (domestically and/or at the summit) or testing positions.

The interplay between summit coverage and summit proceedings may bring about short-term discursive opportunity structures causing actors to adjust communication strategies ad hoc. This was visible in NGOs' partial abandonment of the course of positive, government-friendly rhetoric and launch of more negative attacks in response to Japan's provocative denunciation of the Kyoto Protocol.

NGOs' impromptu strategy adjustment did not lead to desired results, however, as Japan had prescribed itself immunity from any public pressure – likely based on a lesson from a similar situation at the 2007 climate summit in Bali. This illustrates the role of actors' learning effects in shaping summit communication. Generally, NGOs were found to possess more resources for evaluating their summit-based public diplomacy than governments.

8. The summit as world stage?

As final chapter of the study, the following pages serve three purposes: Firstly, the key empirical findings are summarized in reference to the initial research questions. The findings are also assessed in light of my conceptual model and a revised, empirically tested model presented (section 8.1). Secondly, I provide a normative evaluation of my findings and thus relate back to the normative context established earlier. The underlying question is to what extent phenomena investigated here may make up for (or contribute to) the democratic deficit in global governance (section 8.2). Thirdly, I briefly point out some theoretical gains and make a suggestion for how future research could build onto this research (section 8.3).

8.1 Summary of empirical findings

The empirical efforts of this research were guided by both *descriptive* and *explanatory* ambitions. In an initial step, I explored and described the structures and processes as well as strategies and activities of political actors' public communication at the 2010 climate summit in Cancún. The research questions to be answered in this context were 'how' questions; their purpose was to discover the little-explored phenomenon of summit PR in a detailed manner. These explorations and subsequent descriptions were focused on selected aspects provided by my conceptual model or the overall theoretical background, such as actors' on-site routines or their consideration of the event's transnational reach in their communication strategies. This descriptive effort brought about the following key findings:

Conscious use of HIPS as public diplomacy resources

First and foremost, HIPS are indeed consciously exploited as public diplomacy resources, at least in the case of governments with particular ambitions on the international stage (particularly the BRICS countries) as well as NGOs. These actors openly appreciate the high levels of worldwide attention surrounding these events and recognize their value in introducing messages into national public discourses. This exploitation of summits' communicative impact may be motivated by various objectives. On the one hand, summits accommodate decisive negotiations, in which actors hold particular stakes and hence follow specific preferences. They may choose to facilitate their interests regarding these critical moments of policy-making with public communication of policy-relevant nature. In this perspective, HIPS constitute turning points within particular supranational negotiation tracks and are hence appropriate occasions for complementary transnational communication efforts. On the other hand, HIPS can also be seen as global moments *outside* particular negotiation tracks and hence as universal resources for simultaneously speaking to audiences in different countries. Regardless of the particular content of an actor's message – making it public at a summit, where the world's media are assembled, can be regarded as *efficient* approach, as it saves expenditures that would otherwise have to be invested in separate multi-country communication efforts. Brazil, for example, while also being active in public diplomacy outside times of summits, makes particular use of these high-attention episodes. Furthermore, the integration of an actor's message with a summit's symbolic significance might also make messaging more *effective*. This approach could be detected in the case of Mexico, which drew on the severely damaged shape of global climate talks after Copenhagen and the world's cautious curiosity whether the process would ultimately die in Cancún for the purpose of spreading the image of a modern country capable of managing global issues through transparency and inclusivity.

Behind the character of HIPS as resource for cross-border communication looms an explicit agenda by the UNFCCC. While it is formally obliged to show political neutrality and restrict its work to a mere coordination of international policy-making, it does subtly enforce a program of accelerating climate action and promoting the urgency of the issue. This is visible in the organization's various efforts in providing infrastructure that supports political actors in their production of public communication at climate summits and hence promotes worldwide attention to the negotiations. The UNFCCC wants the world to know about the state of the talks, about who is facilitating and who is obstructing agreement. That is why it requires host governments to set up not only the media center, which makes reporting from the summit much easier for journalists, but also to encourage the emergence of non-journalistic links between the summit and the rest of the world. Facilities like the 'Bloggers' Loft' (see 5.2.3) or the streaming of NGO media briefings on summit websites are clear indications for a determination on the side of the UN to make the event resonate with civil society around the globe. While not openly stated, this is informally acknowledged by UN representatives. Also, they do not object to actors exploiting climate summits for other communicative purposes than those imminently linked to the issue of climate change. Host governments' secondary messaging motives, for instance, are commonly taken into account when preparing and carrying out the event.

Variety of strategic communication approaches at the Cancún summit

The particular communication structures and processes as well as strategies and activities that political actors built onto the resource of the Cancún summit varied significantly. More variance was found on the side of governments, where some countries relied on rather traditional, unfocused approaches or individual politicians' commitment, while others carried out more ambitious and professionalized communication operations. Three distinct approaches of summit PR could be identified: (1) *traditional, quiet media work*, (2) *colorful one-man show*, and (3) *professionalized, highly visible PR mix*. NGOs' summit communication appeared less heterogeneous, although contrasting emphases, e.g. on photo ops versus media relations, were found. Here, approaches of summit PR could be lined up on a spectrum between (1) *inventive campaigning* and (2) *educating press*. Section 6.4 summarizes these five different ways of doing public diplomacy at HIPS; tables 6.2 and 6.3 compare the findings on the different dimensions of summit PR.

Differences in how political actors exploited COP-16 for strategic communication purposes became particularly visible in their messaging strategies, i.e. their approaches in devoting summit communication to an overarching idea, theme, or slogan. Here, governments showed again more variance than NGOs; the diversity of governments' messaging ranged from very detailed, policy-related statements that lacked overarching themes to broader, image-related messages that were not connected to climate change. Generally, only few governments invested in the development of a more elaborate message than the usual affirmation that the country would stand up to its responsibility in combating climate change. Especially industrialized countries' governments, like those of Germany or the US, appeared to lack the determination to think more closely about which point they want to bring across at the summit. The largest amount of preparation went into Brazil's messaging, which fused a climate-related but not overly technical message regarding the Amazon region with the overall theme of political and economic progress and was publicized consistently across several channels. In the case of Mexico and South Africa – the current and subsequent host of the COP –, the promotion of these countries' potential as tourist destinations played a major role in messaging. They clearly saw the hosting of the climate summit as a golden opportunity to present to the world a desired country image beyond the specifics of climate negotiations.

On the side of NGOs, only two broader messaging strategies could be found: that of Friends of the Earth and that of the rest of the NGOs in the sample. While the former carried out its conventional messaging approach of highlighting the social dimension of climate change and calling for 'global climate justice' (a message that was commonly described as more 'radical'), the majority of the NGO camp used the Cancún summit for rolling out a new set of messages that were all aimed at lowering the expectations towards the outcomes of the summit and preventing another disappointment in the tradition of Copenhagen (see 6.2.2). This new approach of explicit expectation management and prevention of symbolic overloading was developed under the roof of the NGO alliances, CAN International and GCCA, and then implemented in similar ways by member NGOs. This concerted effort highlights NGOs' tendency to treat summits as valuable moments of worldwide attention whose image can be engineered and integrated into master narratives. For NGOs, global public perception of climate summits resembles modeling clay – it can be sculptured and reshaped according to political needs. How audiences around the world view the summit, what symbolic meanings and political functions they attribute to the event, can – in NGOs' perception – be strongly influenced by their strategic communication decisions. In this effort, they view media as the central tool for setting the tone of the summit as well as intervening in summit negotiations (see 7.1.2).

Coproduction routines between PR professionals and journalists

While a proper assessment of public diplomacy professionals' degree of influence on summit coverage would require an adequate content analysis, there is some first evidence that the communicative offerings provided by PR professionals of governments and especially NGOs are indeed incorporated into the external image of the event as distributed by the media. This is due to a setting of coproduction (see 7.1.1), in which actors' strategic communication efforts and journalists' summit reporting practices are closely intertwined. Both sides are bound through various forms of informal interaction and ad-hoc teamwork and hence craft the image coming out of the summit collaboratively. Particularly significant is the role of NGOs' chief communicators (especially those of Greenpeace International and WWF International), who act as 'information brokers' and provide journalists with guidance and proposals for summit coverage. NGOs' memos and analyses are particularly valuable to journalists if the subject matter to be reported on is of high complexity and the pressure for editorial output is strong (both is usually the case at climate summits). Greenpeace International described its positive rhetoric regarding the outcome of the Cancún summit – which it made available to journalists in form of a succinct analysis distributed in the final summit hours – as strongly influential for media's final reporting. While such claims should always be treated with considerable skepticism, some preliminary reading of media's final assessment of the Cancún summit as well as observations of evident coproduction structures do provide some evidence in their favor. If PR professionals and journalists spend two weeks within the same space, if they cooperatively develop the visual representation of the summit and exchange viewpoints and ideas, it is not unlikely that NGOs' attempts at forming the summit's media image have at least some effect.

Formation of alliances between governments and NGOs

The relationship between PR professionals and journalists is not the only noteworthy constellation involving different types of actors. Interaction among actors of the same type (e.g., between NGOs under the roofs of CAN International and GCCA or between governments as part of a country grouping like BASIC or the Group of 77) has been long known. However, partnerships also occur between NGOs and governments, as was visible in the close contacts that Friends of the Earth and the government of Bolivia (the only country not to support the Cancún Agreements) maintained throughout the summit. Bolivia recommended itself as partner to Friends of the Earth since it called for far more ambitious climate action than other governments and highlighted the responsibility of the industrialized world – much in line with Friends of the Earth's more 'radical' stance. Bolivia provided Friends of the Earth with a negotiation position that the NGO could commonly highlight as ideal. The Bolivian government, in turn, was content to have a reputable civil society partner and thereby nourish its image as the people's representative in the global climate talks, which it had first created by hosting the World People's Conference on Climate Change in Cochabamba in April 2010, conceived as civil society alternative to the failed talks in Copenhagen. At the Cancún summit, Bolivian president Evo Morales paid a visit to Friends of the Earth's delegation, during which both actors (according to FOEI communicators) reinforced their views and emphasized congruencies.

Similar forms of cooperation could be found between NGOs and the small island states of AOSIS before and after the Cancún summit. AOSIS countries are popular partners for NGOs, as their imminent fate illustrates the adverse consequences of climate change in a plain and drastic manner. Already in Copenhagen, NGO campaigners had frequently highlighted the vulnerability of small island states and called upon other countries to 'stand with Tuvalu'. Campaigning for AOSIS countries provides NGOs with an additional moral edge and reduces the complexity of climate change to the unambiguous and forceful image of islands sinking in the ocean. In return, working together with NGOs supports these states in promoting their cause and gives them access to precious PR expertise. NGOs indeed advise AOSIS governments on questions of climate public diplomacy (possibly, the NGO-like stunt of the Maldivian cabinet meeting underwater prior to the Copenhagen summit was inspired by such input) and do not hesitate to run campaigns in explicit conjunction with them, as was illustrated by a joint online campaign of AOSIS and the NGOs Avaaz.org and 350.org during the Copenhagen summit.

Both examples highlight the permeability of separations between actor types. The anticipation of mutual benefits may cause governments and NGOs to work hand in hand and engage in relationships of exchange (a voice at the negotiation table in exchange for public support, a moral edge and forceful message in exchange for PR expertise). In accordance with prototypical conceptions of global governance, climate summits can indeed be identified as forums in which actors of different types flexibly engage in varying configurations, divide tasks and join forces. Their contrasting resources for shaping international policy-making (institutionalized access vs. soft power) can at times turn out to be highly complementary and facilitate the emergence of temporary coalitions in summit discourse encompassing different actor types.

Public diplomacy of truly transnational nature as particular challenge

The extent to which the summit's character as transnational communication resource, i.e. as event that is observed simultaneously across multiple countries, is explicitly considered in political actors' planning varies. While an abstract conception of COPs as targets of worldwide attention is widespread among communicators (see above), processes of reflection and strategizing devoted to making summit communication match the event's *trans*national reach are more uncommon. Out of those actors *not* explicitly dealing with aspects of transnationality in summit communication, most were governments primarily targeting their domestic audiences and therefore not interested in questions of efficacy of multi-country PR. Other actors appeared to have according ambitions but lacked the resources to systematically reflect on those aspects.

Where a specifically transnational outlook in communication strategies could be detected, it became visible in (1) the choice of target outlets and (2) the appropriate crafting of messages and visuals. In terms of target outlets, actors with the explicit ambition of having their communication strategies reach audiences in more than one specific country regard international news agencies like AFP, AP, IPS, or Reuters, Anglo-Saxon papers like *The Financial Times*, *The Guardian*, or *The New York Times*, and eminent networks like Al-Jazeera English, BBC World News, and CNN International as suitable channels. For managers of transnational PR, these outlets are at the top of a global news hierarchy; messages placed with these outlets circulate around the world and trickle down to leading national media and from there further on to specialized outlets. However, none of these outlets should be assumed to have worldwide reach, as communicators pointed out (the market penetration of international news agencies, for instance, differs among continents). Most effective for global messaging would be to target a combination of these outlets.

When it comes to the content communicated via these transnational channels, communicators acknowledge a fundamental challenge, namely to craft messages that resonate widely across national borders. On the one hand, this entails being more specific, i.e. introducing background knowledge that domestic audiences might already hold (like in the case of Brazil's complex Amazon message); on the other hand, messaging of transnational scope should be as blunt and straightforward as possible. This means avoiding all forms of irony, subtlety, or metaphor – as well as any other rhetorical device that is culturally specific. The message should ideally stand for itself, anywhere on Earth. Particularly effective vehicles for such messages are images (see 6.2.6 for a discussion of NGOs' photo ops and their worldwide impact). Here again, aspects of interculturality have to be considered: Does the message become unmistakably clear from the picture? Could the image hurt anyone's feelings? (One government communicator remembered with shock how a contracted PR firm once illustrated a publication for international circulation with photos of half-naked people.) And does the image possess sufficient visual appeal to act as eye-catcher? On this point, communicators commonly draw on representations of widely-known icons and visual clichés, such as the Mayan pyramids of Chichen Itza or spotless Caribbean beaches in the case of the Cancún summit.

However, some communicators appeared skeptical whether truly transnational messaging is possible at all. They highlighted the value of customizing overarching messages to national contexts so as to increase their efficacy. According to their reasoning, audiences confronted with messages that originated beyond their national borders habitually ask 'how does this concern my country?'. Only communication that provides an appropriate answer may pass this first filter. National adaptation should hence be seen as prerequisite for effective communication. Accordingly, one NGO communicator described the staging of spectacular photo ops as the only truly transnational public diplomacy activity.

As mentioned above, this research was additionally motivated by an explanatory motive. Here, the pertaining research questions were 'why' questions; they were directed towards the reconstruction of causal links, or at least relationships of influence, namely at the shaping of political actors' communication strategies through various structural/long-term and dynamic/ad-hoc factors.

These factors were derived from my deductively developed conceptual model, and their applicability and strength of influence was tested empirically. This explanatory effort brought about the following key findings regarding the factors that make political actors choose particular summit communication strategies:

Actor type and international standpoints as most influential factors

Generally, the *type of actor* and its *worldviews and standpoints on the international stage* can be regarded as most influential factors. Whether the actor is a government or an NGO, unsurprisingly, has great implications for its communicative behavior. While governments concentrate their summit resources mostly on policy-making behind the scenes and employ communication in support of this primary function, NGOs are not institutionally guaranteed a seat at the negotiation table and therefore rely on the use of communication to a much greater extent. This has consequences for the sophistication of strategies and diversity of activities carried out by these actors; as natural PR players, NGOs typically show more sophistication and ambition in the various aspects of summit communication than governments (see 6.3.1). However, equally important in shaping summit PR are actors' perceptions of their position within the international system and motivations to induce change to this order. This was particularly evident for the two self-confident emerging economies of Brazil and India, which conducted rather bold, target-driven summit communication – especially if compared to the more traditional, ponderous approaches of recognized players like Germany or the US. Likewise, on the side of NGOs, different preferences regarding how international policy-making should progress also shaped the overall orientation of summit PR. This was observable in the division between more 'radical' NGOs employing rather harsh rhetoric and conducting independent protests and mainstream groups promoting optimistic, more government-friendly messages and coordinating their communication within NGO alliances (see 6.2.3).

Professionalization and professionals' backgrounds as conditional factors

Of less straightforward nature is the influence of the *degree of professionalization* and the *individual background of PR professionals* on actors' choice of communication strategies. While professionalized communication structures and processes – professionalization was understood as incorporation of dedicated PR expertise and exercise of autonomy in taking PR decisions (see 3.3.2) – can strongly shape communication strategies (as in the case of Brazil), my research showed that a complete lack of such foundation can still be counterbalanced by the efforts of an ambitious and talented figurehead (as in the case of India). The

Indian environment minister was no trained PR professional and did not enjoy the support of dedicated communication structures and processes, but he still led his country to similar levels of visibility and recognition as Brazil's sleek PR apparatus. Given this finding, it would be misleading to attribute to this factor a generally dominant role in shaping actors' appearance at the summit; the case of India teaches us otherwise. The degree of professionalization can only make a difference within the limits determined by other factors, like the ambitions on the international stage embodied by the Indian environment minister. Even more dominated by such principal factors is the factor of individual PR professionals' background. While their training and preferences may pose an influence on communication strategies, this happens mostly on the level of fine-tuning, i.e. selecting and reflecting on particular activities (see 6.3.4). The overall orientation of summit communication is determined by other factors.

Short-term discursive opportunity structures as ad-hoc challenges

Finally, on the side of dynamic/ad-hoc factors, the specific unfolding of events at the summit could indeed be detected to pose a short-term influence on political actors' communication strategies. This became particularly evident with regard to Japan's renunciation of a possible extension of the Kyoto Protocol. This move struck many NGOs by utter surprise and forced them to adjust their communication strategies. While most NGOs in the sample had initially set out on a course of positive, not overly hostile messaging (see 6.2.2), this approach had to be revised and was supplemented with efforts to publicly isolate and shame Japan. Thus, the short-term discursive opportunity structure (see 3.3.2) that had emerged from the country's move was utilized by NGOs through suitable modifications to their summit PR. However, in the context of the Japan example, these adjustments did not appear to be very effective. While at COP-13 in Bali, the Japanese government, in a similar situation, had shown responsiveness to public pressure, at the Cancún summit, it prescribed itself complete immunity to any campaigning efforts. From the outset, it highlighted its rigid, nonnegotiable position and shrugged off any international opposition. Therefore, the short-term, summit-internal feedback cycle envisaged in my conceptual model could only be partially confirmed. Summit proceedings as well as their coverage might generate short-term discursive opportunity structures that cause actors to modify their communications strategies ad hoc. But these modifications are not necessarily effective if an actor pursues a course of self-immunization.

Based on these empirical findings, my conceptual model can now be adjusted and specified accordingly (see figure 8.1). In this revised version, it presents all factors that were detected to have an influence on political actors' choice of summit communication strategy. It further pays tribute to the fact that different strengths of influence could be reconstructed as part of my analysis and depicts them as arrows of varying thickness. Bold arrows represent strong, obvious, or otherwise significant influences (e.g. for the factor of worldviews and standpoints or as part of coproduction between PR professionals and journalists), while dotted arrows originate from rather weak or mediating factors. Standard arrows, finally, indicate relationships of influence whose strength appeared to vary across political actors or is of rather general nature and could hence not be qualified more precisely.

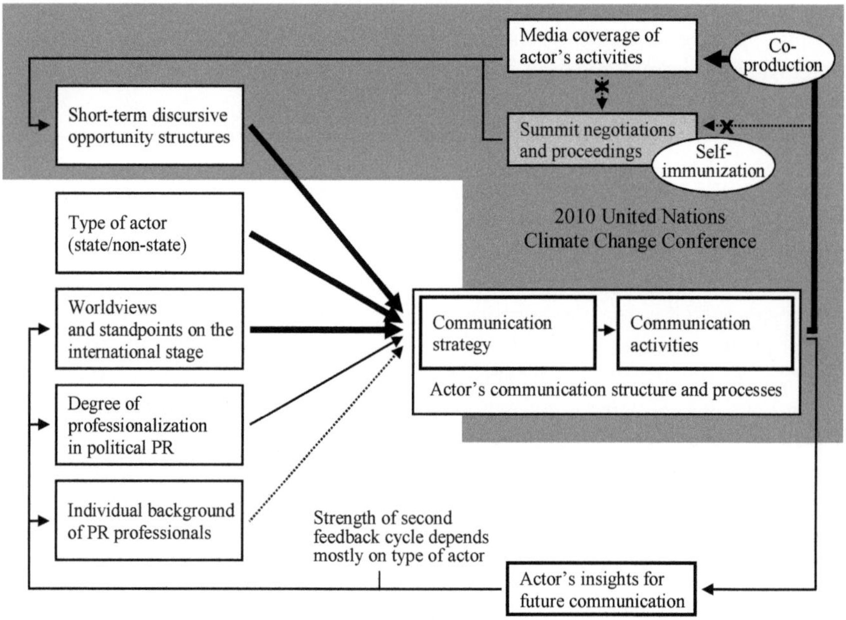

Figure 8.1: Revised conceptual model (after empirical testing)

The conceptual model in this revised version hence provides an empirically verified overview of which relationships of influence shape political actors' communication strategies at HIPS. It has proven to be a helpful instrument for analyzing public communication processes at these large-scale events and is open for further specification based on future research.

8.2 Normative assessment of empirical findings

I now offer a normative assessment of political actors' exploitation of HIPS as resources for public diplomacy. As discussed in the introduction and theoretical part, HIPS can be perceived as triggers of short-term, issue-bound transnational public discourse, which might generate legitimacy for otherwise poorly legitimated institutions of global governance. But does the kind of public communication originating at HIPS and political actors' attempts to influence it really justify the attribution of a legitimacy-generating role to HIPS? How can strategic communication at HIPS be evaluated normatively? Based on pertaining literature and its criteria for evaluating the democratic quality of global governance procedures (see the references in the following), I assess the role of summit communication in the (1) *provision of transparency and access*, the (2) *representation of civil society concerns*, and the facilitation of (3) *governments' responsiveness to civil society concerns*. Additionally, I also evaluate the (4) *shaping of summit coverage by NGOs* through coproduction. The mixed assessment resulting from this evaluation (some aspects can be seen as positive, but there still is substantial need for 'democratization') is summarized in table 8.1 at the end of this section.

Provision of transparency and access

Compared to other negotiations in global governance, UN climate summits appear to provide relatively high levels of transparency and access. This is visible in three features of the event, which all seem to result from deliberate decisions by the UNFCCC and its parties. Firstly, the regime maintains a policy of making relevant documents and records publicly available. Not only are agendas, minutes, and communiqués posted online, the websites of UNFCCC and/or respective host government also feature video recordings of key sessions and media briefings. This effort in opening summit proceedings up to interested publics is, secondly, supplemented by the provision of communication infrastructure supporting journalists as well as representatives of NGO media. While media centers have become an essential component of summits (even of those appearing rather obscure and secretive to the public), the UNFCCC explicitly welcomes media coverage, provides informal acts of support to media representatives wherever needed, and even invests in the development of strategies for maintaining and promoting media interest in the climate negotiations. All of this stems from the underlying belief that media scrutiny facilitates progress in international climate politics.

Part of this plan to make the climate negotiations resonate with publics around the world is also, thirdly, their openness to civil society actors, i.e. NGOs.

As discussed above (see 3.2.3), NGOs fulfilling a set of basic criteria can apply for observer status, which allows them to access the summit venue, participate in selected meetings by making oral interventions, circulate printed materials, and engage in lobbying (Oberthür et al., 2002, pp. 128-134). Consequently, NGOs have become an integral, visible component of climate summits. While the strictness of the rules governing NGO participation can fluctuate from summit to summit (see 5.2.1 and 5.2.2), the average degree of integration of civil society *at the summit itself* – not at protest events outside the confines of the summit venue – is remarkably high in the case of the climate summits. This is corroborated by Fisher (2004), who compared the summits of the UNFCCC with those of IMF and World Bank regarding their openness to NGOs. For both negotiation tracks, she calculated a 'disassociation index', the ratio of the number of protesters *outside* these meetings and that of civil society representatives on the *inside*. While the index goes up to 75 for the financial summits, it remains around 2 for the climate summits (meaning that there are only twice as many protesters in front of the summit gates than observers on the summit grounds). While the underlying data is a bit outdated, these findings, according to the author, point to the value of open-door policies for increasing the legitimacy of international negotiations and, as a consequence, preventing violent protest:

> "By allowing multiple members of organizations to participate inside the halls of the meetings – if not inside the rooms themselves – while at the same time providing a certain level of transparency to their activities, it is likely that there will be less civic dissatisfaction with the process itself. When there are high levels of civic dissatisfaction, it tends to be expressed by transnational social movements in the form of protests – many of which turn violent" (Fisher, 2004, p. 195).

If actors from civil society are admitted to the talks, they can act as "'transmission belt'" (Nanz and Steffek, 2004, p. 323) and "discursive interface" (ibid.) between citizens and institutions of global governance. This works in two directions: NGOs may aggregate citizens' preferences and represent them at the negotiations, i.e. "channel them into the deliberative process of international organizations" (ibid.) and, at the same time, explain the complex substance and process of the negotiations to citizens, i.e. "formulate technical issues in accessible terms" (ibid.). Being allowed to participate in the talks is a basic prerequisite for performing these normatively desirable functions. In the case of the UNFCCC, this condition can be regarded as mostly fulfilled. In my interviews, some NGO representatives even voiced mild surprise about the general degree of openness at COPs – just as one campaigner who unexpectedly found himself in a room full of heads of government at a previous summit.

NGOs as representatives of civil society concerns

However, granting civil society actors access to the negations and thereby facilitating their influencing and deciphering of negotiations on behalf of citizens is only one side of the coin. It is of equal normative importance to ask whom these actors speak for, to what extent they actually represent and are accountable to citizenries worldwide. Accordingly, "democratizing global governance involves more than increasing the number of participants involved in multilateral decision-making processes" (Betsill, 2008b, p. 204); we also have to investigate who profits from this increase in numbers.

With regard to the Cancún summit, a mixed picture of NGO representation, as visible in summit communication, emerges. Most noticeable were the activities by the three most widely known ENGOs (Friends of the Earth, Greenpeace, and WWF) and the two alliances.[14] As was discussed earlier (see 6.2.3), a fundamental split went through this group of NGOs, not only in terms of messaging strategy but also with regard to their underlying ideologies and cultures of representation: On the one side, there were Greenpeace and WWF, whose work around the globe is coordinated by well-resourced international headquarters located in Amsterdam and near Geneva. Both are of the type of organization that Dryzek (1999) describes as mainstream group with "large offices [...], access to the corridors of power, highly paid executives and slick fundraising operations" (p. 47). It is "tightly controlled from its centre" (p. 46) and does not shy away from cooperating with industry (as could be observed for both NGOs in Cancún). The summit communication of these actors is run by educated, white Westerners who possess considerable leeway and resources for determining what their organization says in public and promoting their 'NGO brand'. In Cancún, the work of both NGO alliances, CAN International and GCCA, was closely intertwined with that of these two NGOs.

On the other side, there was Friends of the Earth, which features a rather confederal structure, with individual chapters having a greater say vis-à-vis the small international secretariat. The NGO is critical towards industry cooperation and emphasizes the responsibility of industrialized countries in combating climate change. It considers itself a representative of the Global South and part of the environmental justice movement, which concentrates on the social implications of climate change, especially in developing countries, and is organized "from the bottom up rather than the top down" (p. 47). Instead of coordinating under the roof of the two NGO alliances, Friends of the Earth sees its allies more

14 The pronounced visibility of these five organizations was not just an effect of according sample selection; besides these five players, NGO actors with sustained visibility were indeed rare at COP-16 – with the exception of maybe Oxfam and 350.org (see also section 4.3).

in less institutionalized grassroots groups, like the *Via Campesina* peasant movement, with which it maintained close cooperation in Cancún. The strength of such networks is perceived to be their close ties to groups of citizens and stakeholders that are commonly neglected in formal politics, such as indigenous peoples or other minorities (Beisheim, 2004, pp. 78-79).

It becomes apparent that the summit brought together civil society actors standing for different cultures of representation and accountability: centralized campaigning implemented by Western elites vis-à-vis bottom-up movement-building integrating Southern networks. Problematic from a normative viewpoint is that the former clearly seemed to dominate summit communication: Firstly, the degree of visibility of Friends of the Earth and its messages in Cancún was generally lower than that of Greenpeace and the two alliances, mostly due to a lack of on-site photo ops. Secondly, its capacity to move media coverage into a desired direction appeared to be more limited than that of all other organizations; the NGO appeared to be less integrated in networks of coproduction (see 7.1.1) and admitted to its being overruled by other players in these efforts. Thirdly, according to its own information, the NGO's position was also rejected by all governments except for the 'blocker' of Bolivia, creating a (self-)image of an unwanted, marginalized player.

These observations corroborate long-established criticism of a bias in NGO representation (see Beisheim, 2004; Bestill 2008b for overviews). Already in 1998, Clark, Friedman, and Hochstetler noted that "North-South differences and concomitant differences of philosophy remain a significant source of 'unaligned' NGO frames, or social division" (p. 23). But still at today's climate summits, the loudest voices and closest ties to governments seem to be owned by mainstream NGOs rooted in the industrialized world. These organizations appear to make the most effective use of the access provided by the UNFCCC, as an NGO official once remarked cynically: "'the Africans are watching, the Asians listening, the Latin Americans talking while the North Americans and Europeans were doing business'" (p. 12). Moreover, some of the less institutionalized and more contentious social movement actors of the South (like *Via Campesina*) do not even obtain accreditation at summits but are confined to making their voices heard from the outside (Fisher, 2004, p. 182).

The UNFCCC commonly takes small and informal, yet nonetheless noteworthy steps in offsetting this normative deficiency of a bias in NGOs' summit communication by empowering more inexperienced actors. This can be seen in unofficial acts of support, such as guidance on how to write a press release, making use of UNFCCC contacts to journalists, or even organizing a visit by the Executive Secretary to increase the visibility of smaller NGOs' activities. The UNFCCC calls this 'capacity-building', and despite their ad-hoc and limited nature, these efforts can be seen as normatively valuable. They generally reflect the progressive attitude of the organization in promoting the importance of the issue and facilitating political action. This outlook can also be detected in UN-FCCC-organized and foundation-sponsored programs bringing journalists from developing countries to the summits, where they are trained in the complex substance matter and supported in covering the event for their domestic audiences. Through both approaches, the UNFCCC contributes to the empowerment of disadvantaged – and hence less visible – stakeholders.

Responsiveness of governments to civil society concerns

Questions of access and representation are about the preconditions for influencing governmental decisions in institutions of global governance (Nanz and Steffek, 2004, p. 323). The fundamental question is to what extent the concerns and demands voiced by civil society (be it inside the summit, in front of its gates, or via the media) are adhered to by governments and "reflected in the decision-making process of international organizations" (ibid.). If governments show sufficient degrees of responsiveness towards civil society input and candidly consider these contributions in their negotiations, NGO involvement could be seen as a normatively valuable surrogate for the abstract, long, or simply nonexistent mechanisms of legitimization in global governance (Beisheim, 2004, pp. 61). This would still be a far cry from the "deliberative forums" or "participatory arenas" that Nanz and Steffek (2004, p. 322) envision at the global level, which would possess "the prerogatives to scrutinize and monitor policy choices on international organizations" (ibid.) and "introduce a deliberative element to the public level" (ibid.). But it would be a start.

My normatively assessable findings regarding this aspect are mixed. There were governments in my sample that explicitly appreciated the exchange with and contributions by NGOs, particularly Brazil, India, and South Africa, some of which even let NGO representatives come to Cancún on an official delegation ticket. However, considering the fact that I researched government *PR*, it is hard to assess whether this appreciation of NGO input was merely motivated by strategic ambitions (such as the promotion of a particular image within the environmental community) or backed up by serious political interest in civil society positions. Given the fact that NGOs do not like to be exploited for 'greenwashing', we can assume that rhetoric was at last partially turned into action. As an exception, the host government of Mexico showed some skepticism regarding the value of NGO participation, since it was also perceived as a potential source of disruption of smooth summit proceedings.

Harder to find than these general acknowledgments of the important role of NGOs were explicit considerations of their demands – let alone visible policy changes made in response. NGOs may be assumed to hold 'communicative power' (see section 3.1), and only "a cynic might see this as capable of generating only a few pinpricks in the hides of established powers" (Dryzek, 1999, p. 45), but the truth is that NGOs' "capacity to affect the terms of discourse" (ibid., p. 46) was rather limited at the Cancún summit. If we think back to Japan's renunciation of a Kyoto Protocol extension – *the* dominating issue of the summit, which triggered a lot of attention and strategic NGO communication (see section 7.2) –, there was no evidence for responsiveness to civil society concerns on the side of the Japanese government. Quite the opposite, the Japanese followed a path of total abstention from what Risse (2000) describes as 'arguing', which, in the setting of international negotiations, may have a "'civilizing' effect" (p. 22) on actors and unlocks "the power of the better argument" (p. 20). (To be fair, also on the side of NGOs, parts of communication around the issue lacked the argumentative value posited by Risse's Habermas-inspired conception of public discourse.) Apparently, Japan had learned from COP-13 in Bali, where the country's unclear position, combined with public pressure, did in fact result in governmental responsiveness. And as Risse theorizes, actors are indeed more likely to engage in arguing if they are uncertain about their own interests and the 'rules of the game' and eager to engage in actual problem-solving (pp. 20-21). Through its explicit self-immunization, Japan avoided the reemergence of this constellation, which it knew from Bali, from the outset.

Generally, the hardened fronts at climate summits should not be regarded as overly conducive to a setting in which proper arguing can take place and NGOs can unleash the power of the better argument onto governments. However, whenever a government is unclear about its position or aims at promoting its standing with civil society (like the BRICS governments mentioned above), some degrees of listening and responding to NGOs may be found. As part of their 'spiral model' (see 3.1.3), Risse and Sikkink (1999) describe – albeit for a different setting – how governments that are at least minimally concerned about their international image will always respond and maybe even make small concession in response to international criticism. This in turn opens the door for further interventions by NGOs, and slowly a dynamic of arguing and gradual change is set in motion. In other words, communicative power employed by NGOs (or other actors) causes targeted governments to tie themselves to an argumentative position, which then triggers counter-arguments by NGOs, which in turn brings about argumentative repositioning on the side of governments, and so forth. It is surely possible that this process of enhancing governmental responsiveness to civil society concerns does also take place at HIPS – provided that a targeted government does not follow the path of explicit self-immunization.

Shaping of summit coverage by NGOs

Finally, how are we to assess coproduction among PR professionals (especially those of NGOs) and journalists (see 7.1.1)? Taken at face value, a too intimate relationship between both sides endangers the independence of the press and leads to biased coverage. Journalists should instead show critical distance to PR offerings and stay in control of the entire editorial process. Nonetheless, there are two arguments why the coproduction detected in Cancún should not be regarded as too much of a normative problem – one empirical and one political in nature.

Provision of transparency and access	**NGOs as representatives of citizens'/ stakeholders' concerns**
- Online publication of documents and videos of selected sessions - Infrastructure and support for journalists and 'NGO media' - Accreditation of NGOs as observers with multiple privileges	- Division in ENGO community (between mainstream/'Northern' and environmental justice/'Southern' groups) not reflected in visibility, media influence, and access to governments - Sporadic acts of unofficial support to disadvantaged actors by the UNFCCC
Positive: Relatively high degree of access and transparency, especially compared to other negotiation summits	*Need for improvement:* Dominance of selected civil society views; no institutionalized mechanisms for offsetting this bias
Integration of citizens'/ stakeholders' concerns by governments	**Shaping of summit coverage by NGOs**
- Rhetorical appreciation of the role of NGOs by most governments - Low impact of NGOs' 'communicative power' on the Japan issue, due to Japan's self-immunization and explicit abstention from 'arguing'	- Coproduction between PR professionals and journalists is equal to manipulation - Strong NGO influence on coverage increases 'communicative power', although not all NGOs are equally integrated in coproduction routines
Need for improvement: Climate summits certainly no 'deliberative forums' or 'participatory arenas'; institutional mechanisms needed for this	*Positive:* Journalists' orientation towards progressive actors facilitates political action; however, some actors lack media relations capacity

Table 8.1: Summary of normative assessment of empirical findings

Firstly, coproduction should not be mistaken for journalists' complete surrendering of editorial control to PR specialists. Quite the opposite, coproduction, as observed at COP-16, means that pertinent decisions are taken collaboratively, through quick, informal interactions, but without an overruling of either side. While journalists might make use of NGO resources for some parts of the editorial process (like the generation of story ideas or the acquirement of background expertise), this does not imply that their (ideally) critical, journalistic mindset is automatically turned off and that they are stripped of any agency or veto power.

Secondly, from a progressive perspective, it might be considered positive if summit coverage includes NGO contributions to an unusually high degree. It thereby becomes an important channel for NGOs' putting pressure on states (see 7.1.2.3) and exercising the 'communicative power' described above. However, as pointed out above, not all part of civil society actors enjoy equal access to the media; some actors are more integrated in coproduction routines than others – not to mention those parts of civil society not represented at the summit. Hence, normative problems relating to NGO representation remain.

8.3 Concluding theoretical remarks

The title of this study promised an investigation into the 'engineering' of 'global public discourse'. But to what extent did it deliver on that promise? Let us start with the latter notion: I assumed that there are certain types of meetings between political decision-makers, namely high-level international political summits (HIPS) that carry the potential of transnationalizing public discourse in several national public spheres at the same time. These events can hence be seen as triggers of short-term transnational public spheres, a mechanism that is significantly facilitated by their symbolic loading and openness to civil society actors.

To be clear, the number of summits that can be counted as HIPS in this strict sense is limited. Not all international gatherings of political leaders are loaded with symbolic resources by organizers, civil society, or media; their drawing of attention on a wider scale and across national borders is more the exception than the rule (which can also be seen in the fact that the summit in focus here was loaded with much less symbolic significance and observed around the world to a much smaller degree than other summits in the same negotiation track). Also, not all political meetings provide the high level of access to civil society representatives that was detected for the climate summits (even though this does not automatically solve all democratic deficiencies in global environmental governance, as was shown in the preceding section). There are summits from which NGOs are banned altogether and that may hence feature fundamentally different conditions for communication processes. Hence, the findings presented here only apply to a specific phenomenon defined by narrow parameters; other fora playing a role in transnationalization processes require other research strategies.

Complex networks of summit public diplomacy

With regard to the notion of 'engineering', this research has shown that HIPS are indeed perceived as communication resources by political actors. Worldwide attention to these events means that governments and NGOs are presented with an opportunity – but, in some cases, also the obligation – to engage in public communication. However, even though the term 'transnational' figures prominently in this study's title, not all strategic communication happening around summits is of transnational nature. As summarized above, genuinely *trans*national public diplomacy, i.e. those communicative efforts that are explicitly crafted for being effective across multiple countries and hence lack national specificities, is rare. Instead – and this is a promising point for further research –, HIPS feature complex public diplomacy constellations that interweave transnational, international, and also solely domestic processes. In the theoretical part of this study, we have come across such notions as 'two-level gaming' the 'boomerang pattern', or the importance of influencing national actors for initiating supranational political change. Empirically, these perspectives have proven to possess much validity with regard to how political actors conduct public diplomacy at HIPS: Governments, for instance, send differing, yet related messages to national and international audiences. NGOs enhance transnationally oriented photo ops with domestic media relations. National politicians enter strategic alliances with transnational movements for attaining national goals. NGO networks specifically refrain from building up domestic pressure on stubborn governments but rope in other governments for campaign purposes. Host countries, finally, make use of worldwide attention for modifying their international image and simultaneously stimulating domestic political change.

All this suggests the conclusion that public diplomacy around summits in global governance is seldom just about country A influencing country B (*international* public diplomacy) or NGO X using a summit for addressing the entire world (*transnational* public diplomacy). Instead, summit public diplomacy brings together various cross-linked processes of transnationally, internationally, and nationally oriented strategic communication that are combined in unique constellations according to actors' strategic objectives (the factors that shape the building of these communication strategies were explored in this study). Hence, drawing on my assumption that HIPS carry transnationalizing capacity for public discourse around the world and the empirical observation that political actors make indeed use of this capacity for strategic communication purposes, future research should investigate these multifaceted networks of summit public diplomacy more closely. While previous research has brought about some insights regarding NGOs' communicative exploitation of these events, the amount of empirical studies on governments' efforts in this regard – as a matter of fact, on governments' public diplomacy *in general* – is surprisingly limited. Communication studies should hence devote more attention to the role of states and NGOs as PR players on the international stage (on the occasion of and outside summits) – the political communication of global governance deserves more elaborate investigation.

Literature

Adolphsen, Manuel & Lück, Julia (2012). Non-routine interactions behind the scenes of a global media event: How journalists and political PR professionals co-produced the 2010 UN climate conference in Cancún. In Hartmut Wessler & Stefanie Averbeck-Lietz (eds.), *Grenzüberschreitende Medienkommunikation – Medien & Kommunikations-wissenschaft: Sonderband 2* (pp. 141-158). Baden-Baden: Nomos.

Agre, Philip E. (2002). Real-time politics: The Internet and the political process. *The Information Society,* 18(5), pp. 311-331.

Alexander, Jeffrey C. & Ronald N. Jacobs (1998). Mass communication, ritual and civil society. In Tamar Liebes & James Curran (eds.), *Media, ritual and identity* (pp. 23-41). London: Routledge.

Anderson, Alison (2009). Media, politics, and climate change: Towards a new research agenda. *Sociology Compass*, *3*(2), pp. 166-182.

Andonova, Liliana B., Betsill, Michele M. & Bulkeley, H. (2009). Transnational climate governance. *Global Environmental Politics*, *9*(2), pp. 52-73.

Asen, Robert & Brouwer, Daniel C. (2001). Introduction: Reconfigurations of the public sphere. In Robert Asen and Daniel C. Brouwer (eds.), *Counterpublics and the state* (pp. 1-32). Albany, NY: State University of New York Press.

Ayish, Muhammad I. (2008). *The new Arab public sphere*. Berlin: Frank & Timme.

Beck, Ulrich (1997). *The reinvention of politics: Rethinking modernity in the global social order*. Cambridge: Polity Press.

Beck, Ulrich (2007). *Weltrisikogesellschaft: Auf der Suche nach der verlorenen Sicherheit*. Frankfurt a. M.: Suhrkamp.

Beisheim, Marianne (2004). *Fit für Global Governance? Transnationale Interessengruppenaktivitäten als Demokratisierungspotential – am Beispiel Klimapolitik*. Opladen: Leske + Budrich.

Bentele, Günter, Liebert, Tobias & Seeling, Stefan (1997). Von der Determination zur Intereffikation: Ein integriertes Modell zum Verhältnis von Public Relations und Journalismus. In Günter Bentele & Michael Haller (eds.), *Aktuelle Entstehung von Öffentlichkeit: Akteure-Strukturen-Veränderungen* (pp. 225-250). Konstanz: UVK.

Berridge, Geoff R. (1995). *Diplomacy: Theory and practice*. London: Practice Hall.

Betsill, Michele M. (2008a). Environmental NGOs and the Kyoto Protocol negotiations: 1995 to 1997. In Michele M. Betsill & Elisabeth Corell (eds.), *NGO diplomacy: The influence of nongovernmental organizations in international environmental negotiations* (pp. 43-66). Cambridge: MA: MIT Press.

Betsill, Michele M. (2008b). Reflections on the analytical framework and NGO diplomacy. In Michele M. Betsill & Elisabeth Corell (eds.), *NGO diplomacy: The influence of nongovernmental organizations in international environmental negotiations* (pp. 177-206). Cambridge: MA: MIT Press.

Betsill, Michele M. & Corell, Elisabeth (2008a). Introduction to NGO diplomacy. In Michele M. Betsill & Elisabeth Corell (eds.), *NGO diplomacy: The influence of nongovernmental organizations in international environmental negotiations* (pp. 1-17). Cambridge: MA: MIT Press.

Betsill, Michele M. & Corell, Elisabeth (eds.) (2008b). *NGO diplomacy: The influence of nongovernmental organizations in international environmental negotiations*. Cambridge, MA: MIT Press.

Beyeler, Michelle & Kriesi, Hanspeter (2005). Transnational protest and the public sphere. *Mobilization, 10*(1), pp. 95-109.

Billett, Simon (2010). Dividing climate change: Global warming in the Indian mass media. *Climatic Change, 99*(1-2), pp. 1-16.

Black, Richard (2010, December 10). Japan targeted on Kyoto climate stance at Cancun summit. *BBC News* (online edition). Retrieved November 7, 2011 from http://www.bbc.co.uk/news/science-environment-11966710.

Blumler, Jay G. & Kavanagh, Dennis (1999). The third age of political communication: Influences and features. *Political Communication, 16*(3), pp. 209-230.

Bogner, Alexander, Littig, Beate & Menz, Wolfgang (2009). *Experteninterviews: Theorien, Methoden, Anwendungsfelder*. Wiesbaden.

Boykoff, Jules (2012). US media coverage of the Cancún climate change conference. *PS: Political Science & Politics, 45*(2), pp. 251-258.

Boykoff, Maxwell T. (2007). Flogging a dead norm? Newspaper coverage of anthropogenic climate change in the United States and United Kingdom from 2003 to 2006. *Area, 39*(2), pp. 470-481.

Boykoff, Maxwell T. (2010). Indian representations of climate change in a threatened journalistic ecosystem. *Climatic Change*, 99(1-2), pp. 17-25.

Boykoff, Maxwell T. & Boykoff, Jules M. (2004). Balance as bias: Global warming and the US prestige press. *Global Environmental Change, 14*(2), pp. 125-136.

Braman, Sandra (1996). Interpenetrated globalization: Scaling, power, and the public sphere. In Sandra Braman & Annabelle Sreberny-Mohammadi (eds.), *Globalization, communication and transnational civil society* (pp. 21-36). Cresskill, NJ: Hampton Press.

Brasil.gov.br (n.d.) *Espaço Brasil*. Retrieved December 12, 2010 from http://www.brasil.gov.br/cop-english/brazil-space/espaco-brasil.

Brosius, Hans-Bernd, Koschel, Friederike & Haas, Alexander (2008). *Methoden der empirischen Kommunikationsforschung: Eine Einführung*. Wiesbaden: VS.

Brossard, Dominique, Shanahan, James & McComas, Katherine (2004). Are issue-cycles culturally constructed? A comparison of French and American coverage of global climate change. *Mass Communication and Society*, *7*(3), pp. 359-377.

Brüggemann, Michael (2008). *Europäische Öffentlichkeit durch Öffentlichkeitsarbeit? Die Informationspolitik der Europäischen Kommission*. Wiesbaden: VS.

Brüggemann, Michael, Hepp, Andreas, Kleinen-von Königslow, Katharina & Wessler, Hartmut (2009). Transnationale Öffentlichkeit in Europa: Forschungsstand und Perspektiven. *Publizistik, 54*(3), pp. 391-414.

Brüggemann, Michael & Schulz-Forberg, Hagen (2009). Becoming pan-European? Transnational media and the European public sphere. *International Communication Gazette, 71*(8), pp. 693-712.

Cammaerts, Bart & Van Audenhove, Leo (2005). Online political debate, unbounded citizenship, and the problematic notion of a transnational public sphere. *Political Communication, 22*(2), pp. 179-196.

CAN International (2010, November 30). *Japan wins 1st (and only) place Fossil of the Day for trying to kill Kyoto Protocol*. Retrieved November 7, 2011 from http://www.climatenetwork.org/sites/default/files/Fossil_of_the_Day_-_Cancun_-_Nov_30_2010_0.pdf.

CAN International (2011). *People*. Retrieved November 7, 2011 from http://www.climatenetwork.org/about/people

Cañizález, Andrés & Lugo-Ocando, Jairo (2008). Beyond national media-systems: A medium for Latin America and the struggle for integration. In Jairo Lugo-Ocando (ed.), *The media in Latin America* (pp. 211-228). Maidenhead: Open University Press.

Canwest News Service (2007, December 14). *Canada targeted in environmental ad.* Retrieved November 7, 2011 from http://www.canada.com/story.html?id=2065c94c-cfe2-4e0e-a06b-ae61b7caceac.

Carey, James W. (2009). *Communication as culture: Essays on media and society.* New York: Routledge.

Carpenter, Chad (2001). Business, green groups and the media: The role of non-governmental organizations in the climate change debate. *International Affairs, 77*(2), pp. 313-328.

Carvalho, Anabela (2007). Ideological cultures and media discourses on scientific knowledge: Re-reading news on climate change. *Public Understanding of Science, 16*(2), pp. 223-243.

Castells, Manuel (1996). *The rise of the network society.* Oxford: Blackwell Publishers.

Castells, Manuel (2008). The new public sphere: Global civil society, communication networks, and global governance. *The ANNALS of the American Academy of Political and Social Science, 616*, pp. 78-93.

Center for Educational Design and Communication (2007, December 13). *Bali (Titanic parody ad).* Retrieved November 7, 2011 from http://cedc.org/design/bali-titanic-parody-ad.

Clark, Ann Marie, Friedman, Elisabeth J. & Hochstetler, Kathryn (1998). The sovereign limits of global civil society: A comparison of NGO participation in UN world conferences on the environment, human rights, and women. *World Politics, 51*(1), pp. 1-35.

COP-16 (n.d. a) *COP-16/CMP-6, Cancún, Mexico.* Retrieved December 14, 2010 from http://cc2010.mx/en.

COP-16 (n.d. b). *Side events and exhibits.* Retrieved November 7, 2011 from http://cc2010.mx/en/participants/side-events-and-exhibits.

Corell, Elisabeth & Betsill, Michele M. (2008). Analytical framework: Assessing the influence of NGO diplomats. In Michele M. Betsill & Elisabeth Corell (eds.), *NGO diplomacy: The influence of nongovernmental organizations in international environmental negotiations* (pp. 19-42). Cambridge: MA: MIT Press.

Costa, Natalia. (2010, February 11). Brazil strives to promote international image. *China Radio International* (online edition). Retrieved October 29, 2010 from http://english.cri.cn/6966/2010/02/11/1361s549800.htm.

Cottle, Simon (2006). Mediatized rituals: Beyond manufacturing consent. *Media, Culture & Society, 28*(2), pp. 411-432.

Cottle, Simon & Rai, Mugdha (2008). Global 24/7 news providers: Emissaries of global dominance or global public sphere? *Global Media and Communication, 4*(2), pp. 157 181.

Couldry, Nick & Dreher, Tanja (2007). Globalization and the public sphere: Exploring the space of community media in Sydney. *Global Media and Communication, 3*(79), pp. 79-100.

Couldry, Nick & Rothenbuhler, Eric W. (2007). Simon Cottle on 'mediatized rituals': A response. *Media, Culture & Society, 29*(4), pp. 691-695.

Cowan, Geoffrey & Cull, Nicholas J. (2008). Public diplomacy in a changing world. *The ANNALS of the American Academy of Political and Social Science, 616*, pp. 6-8.

Cull, Nicholas (2008). Public diplomacy: Taxonomies and histories. *The ANNALS of the American Academy of Political and Social Science, 616*, pp. 31-54.

Daiwa Foundation (n.d.). Climate *and energy policy in post-crisis Japan and the UK.* Retrieved November 7, 2011 from www.dajf.org.uk/annual-seminar/climate-and-energy-policy-in-post-crisis-japan-and-the-uk.

Dayan, Daniel (2010). Beyond media events: Disenchantment, derailment, disruption. In Nick Couldry, Andreas Hepp & Friedrich Krotz (eds.), *Media events in a global age* (pp. 23-31). London: Routledge.

Dayan, Daniel & Katz, Elihu (1992). Media events: The live broadcasting of history. Cambridge, MA: Harvard University Press.
de Jong, Wilma (2005). The power and limits of media-based international oppositional politics – a case study: The Brent Spar conflict. In Wilma de Jong, Martin Shaw & Stammers, Neil (eds.), *Global activism, global media* (pp. 110-124). London: Pluto Press.
Dearth, Douglas H. (2002). Shaping the 'information space'. *Journal of Information Warfare, 1*(3), pp. 1-15.
Delicath, John W. & DeLuca, Kevin M. (2003). Image events, the public sphere, and argumentative practice: The case of radical environmental groups. *Argumentation, 17*(3), pp. 315-333.
della Porta, Donatella & Tarrow, Sidney (2005). Transnational processes and activism: An introduction. In Donatella della Porta & Sidney Tarrow (eds.), *Transnational protest and global activism* (pp. 1-17). Lanham, MD: Rowman & Littlefield.
DeLuca, Kevin M. (2001). *Image politics: The new rhetoric of environmental activism*. New York: Guildford Press.
Deutsch, Karl W. (1953). *Nationalism and social communication: An inquiry into the foundations of nationality*. Cambridge, MA: MIT Press.
Deutsch, Karl W. (1956). Shifts in the balance of communication flows: A problem of measurement in international relations. *The Public Opinion Quarterly, 20*(1), pp. 143-160.
Diani, Mario & Donati, Paolo R. (1999). Organisational change in Western European environmental groups: A framework for analysis. *Environmental Politics, 8*(1), pp. 13-34.
Diekmann, Andreas (2007). *Empirische Sozialforschung: Grundlagen – Methoden – Anwendungen*. Reinbek bei Hamburg: Rowohlt Taschenbuch.
Dirikx, Astrid & Gelders, Dave (2008). Newspaper communication on global warming: Different approaches in the US and the EU? In Anabela Carvalho (ed.), *Communicating climate change: Discourses, mediations and perceptions* (pp. 98-109). Braga, Portugal: Centro de Estudos de Comunicação e Sociedade, Universidade do Minho. Retrieved April 7, 2012 from http://www.lasics.uminho.pt/ojs/index.php/climate_change/article/ view/416/386.
Dirikx, Astrid & Gelders, Dave (2009). Global warming through the same lens: An explorative framing study in Dutch and French newspapers. In Tammy Boyce & Justin Lewis (eds.), *Climate change and the media* (pp. 200-210). New York: Peter Lang Publishing.
Dirikx, Astrid & Gelders, Dave (2010). To frame is to explain: A deductive frame-analysis of Dutch and French climate change coverage during the annual UN Conferences of the Parties. *Public Understanding of Science, 19*(6), pp. 732-742.
Donges, Patrick & Imhof, Kurt (2005). Öffentlichkeit im Wandel. In Bonfadelli, Heinz, Jarren, Otfried & Siegert, Gabriele (eds.), *Einführung in die Publizistikwissenschaft* (pp. 147-177). Bern: UTB.
Doyle, Julie (2009). Climate action and environmental activism: The role of environmental NGOs and grassroots movements in the global politics of climate change. In Tammy Boyce & Justin Lewis (eds.), *Climate change and the media* (pp. 103-116). New York: Peter Lang Publishing.
Drieschner, Frank (2010, December 2). Frieren in Cancún. *ZEIT Online*. Retrieved November 7, 2011 from http://www.zeit.de/2010/49/Klimagipfel-Cancun.
Dryzek, John S. (1999). Transnational democracy. *The Journal of Political Philosophy, 7*(1), pp. 30-51.
Dunn, David H. (2004). The lure of summitry: International dialogue at the highest level. In Christer Jönsson & Richard Langhorne (eds.), *Diplomacy – volume 3* (pp. 137-169). London: Sage.
Eckstein, Harry (1992). Case study and theory in political science. In Harry Eckstein, *Regarding politics: Essays on political theory, stability and change* (pp. 117-176). Berkeley, CA: University of California Press.

Eco (2010, December 1). *Cancun personals.* Retrieved November 7, 2011 from http://climatenetwork.org/sites/default/files/ECO_3_COP_16_Eglish_version_.pdf.
Edwards, Paul N. (2001). Representing the global atmosphere: Computer models, data, and knowledge about climate change. In Clark A. Miller and Paul N. Edwards (eds.), *Changing the atmosphere: Expert knowledge and environmental governance* (pp. 31-65). Cambridge, MA: MIT Press.
Eide, Elisabeth & Kunelius, Risto (2010). Domesticating global moments: A transnational study on the coverage of the Bali and the Copenhagen Climate Summits. In Eide, Elisabeth, Kunelius, Risto & Kumpu, Ville (eds.), *Global climate, local journalisms: A transnational study of how media make sense of climate summits* (pp. 11-50). Bochum: Projektverlag.
Eide, Elisabeth, Kunelius, Risto & Kumpu, Ville (eds.) (2010). *Global climate, local journalisms: A transnational study of how media make sense of climate summits.* Bochum: Projektverlag.
Entman, Robert M. (2008). Theorizing mediated public diplomacy: The U.S. case. *The International Journal of Press/Politics, 13*(2), pp. 87-102.
Esser, Hartmut (1999). *Soziologie: Spezielle Grundlagen – Band 1: Situationslogik und Handeln.* Frankfurt am Main: Campus.
Facebook (2012). *Key Facts.* Retrieved April 7, 2012 from http://newsroom.fb.com/content/default.aspx?NewsAreaId=22.
Fazal, Shehina & Tsagarousianou (2002). Diasporic communication: Transnational cultural practices and communicative spaces. *Javnost – The Public, 9*(1), pp. 5-18.
Ferree, Myra Marx, Gamson, William Anthony, Gerhards, Jürgen & Rucht, Dieter (2002). *Shaping abortion discourse: Democracy and the public sphere in Germany and the United States.* Cambridge: University Press.
Flick, Uwe (2006). *An introduction to qualitative research.* London: Sage.
Flickr (n.d). *Asahi Shimbun - Jan 6th 2008.* Image in an Avaaz gallery. Retrieved November 7, 2011 from http://www.flickr.com/photos/avaaz/2232597563.
Fisher, Ali & Bröckerhoff, Aurélie (2008). *Options for influence: Global campaigns of persuasion in the new world of public diplomacy.* London: Counterpoint.
Fisher, Dana R. (2004). Civil society protest and participation: Civic engagement within the multilateral governance regime. In Norichika Kanie & Peter M. Haas (eds.), *Emerging forces in environmental governance* (pp. 176-199). Tokyo: United Nations University Press.
Fomerand, Jacques (1996). UN conferences: Media events of genuine diplomacy? *Global Governance, 2*(3), pp. 361-375.
Fraser, Nancy (2007). Transnationalizing the public sphere: On the legitimacy and efficacy of public opinion in a post-Westphalian world. *Theory, Culture & Society, 24*(4), pp. 7-30.
Friends of the Earth International (2010, December 1). *Japan threatens progress at climate talks in Cancun.* Retrieved November 7, 2011 from http://www.foei.org/en/media/archive/2010/japan-threatens-progress-at-climate-talks-in-cancun.
Früh, Werner (2007). *Inhaltsanalyse.* Konstanz: UVK (UTB).
George, Alexander L. & Bennett, Andrew (1997). *Case studies and theory development in the social sciences.* Cambridge, MA: MIT Press.
Gerhards, Jürgen (1993). *Neue Konfliktlinien in der Mobilisierung öffentlicher Meinung: eine Fallstudie.* Opladen. Westdeutscher Verlag.
Gerhards, Jürgen (1998). Öffentlichkeit. In Otfried Jarren, Ulrich Sarcinelli & Ulrich Saxer (eds.), *Politische Kommunikation in der demokratischen Gesellschaft: Ein Handbuch mit Lexikonteil* (pp. 268-274). Opladen: Westdeutscher Verlag.
Gerhards, Jürgen & Neidhardt, Friedhelm (1991). Strukturen und Funktionen moderner Öffentlichkeit: Fragestellungen und Ansätze. In Stefan Müller-Doohm & Klaus Neumann-Braun (eds.), *Öffentlichkeit – Kultur – Massenkommunikation: Beiträge zur Medien- und Kommu-*

nikationssoziologie (pp. 31-89). Oldenburg: Bibliotheks- und Informationssystem der Universität Oldenburg.

Gerring, John (2001). *Social science methodology: A criterial framework*. Cambridge: University Press.

Gerring, John (2004). What is a case study and what is it good for? *American Political Science Review*, *98*(2), 341-354.

Gestrich, Andreas (2006). The public sphere and the Habermas debate. *German History*, *24*(3), pp. 413-430.

Giffard, Anthony C. & Rivenburgh, Nancy K. (2000). News agencies, national images, and global media events. *Journalism & Mass Communication Quarterly*, *77*(1), pp. 8-21.

Gilboa, Eytan (2001). Diplomacy in the media age: Three models of uses and effects. *Diplomacy & Statecraft*, *12*(2), pp. 1-28.

Gilboa, Eytan (2008). Searching for a theory of public diplomacy. *The ANNALS of the American Academy of Political and Social Science*, *616*, pp. 55-77.

Gläser, Jochen & Laudel, Grit (2009). *Experteninterviews und qualitative Inhaltsanalyse*. Wiesbaden: VS.

Government of the United Mexican States & Secretariat of the UNFCCC and its Kyoto Protocol (2010). *Agreement regarding the 16th session of the COP to the UNFCCC*. Bonn, Germany. Retrieved September 27, 2010 from http://cc2010.mx.

Gough, Clair & Shackley, Simon (2001). The respectable politics of climate change: The epistemic communities and NGOs. *International Affairs*, *77*(2), pp. 329-345.

Greenberg, Donald W. (1985). Staging media events to achieve legitimacy. A case study of Britain's Friends of the Earth. *Political Communication and Persuasion*, *2*(4), pp. 347-362.

Gregory, Bruce (2008). Public diplomacy: Sunrise of an academic field. *The ANNALS of the American Academy of Political and Social Science*, *616*, pp. 274-290.

Gripsrud, Jostein, Moe, Hallvard, Molander, Anders & Murdock, Graham (eds.) (2010). *The idea of the public sphere: A reader*. Lanham, MD: Lexington Books.

Grunig, J. E. & Hunt, T. (1984). *Managing Public Relations*. New York: Thomson Learning.

Habermas, Jürgen (1979). What is universal pragmatics? In Jürgen Habermas, *Communication and the evolution of society* (pp. 1-68, 208-219). Boston. Beacon Press.

Habermas, Jürgen (1981). *Theorie des kommunikativen Handelns (Band 1): Handlungsrationalität und gesellschaftliche Rationalisierung*. Frankfurt am Main: Suhrkamp.

Habermas, Jürgen (1985). *The Theory of Communicative Action (Volume 1): Reason and the Rationalization of Society*. Boston, MA: Beacon Press.

Habermas, Jürgen (1990). *Strukturwandel der Öffentlichkeit: Untersuchungen zu einer Kategorie der bürgerlichen Gesellschaft*. Frankfurt am Main: Suhrkamp.

Habermas, Jürgen (1992). *Faktizität und Geltung: Beiträge zur Diskurstheorie des Rechts und des demokratischen Rechtsstaats*. Frankfurt am Main: Suhrkamp.

Habermans, Jürgen (1998). *Die postnationale Konstellation: Politische Essays*. Frankfurt am Main: Suhrkamp.

Habermas, Jürgen (2001a). *The postnational constellation: Political essays*. Cambridge, MA: MIT Press.

Habermas, Jürgen (2001b, June 28). Warum braucht Europa eine Verfassung? *DIE ZEIT*. Retrieved April 2, 2012 from http://pdf.zeit.de/2001/27/200127_verfassung.xml.pdf.

Hallahan, Kirk, Holtzhausen, Derina, van Ruler, Betteke, Verčič, Dejan & Sriramesh, Krishnamurthy (2007). Defining strategic communication. *International Journal of Strategic Communication*, *1*(1), pp. 3-35.

Hallin, Daniel C. & Mancini, Paolo (1992). The summit as media event: The Reagan/Gorbachev meetings on U.S., Italian, and Soviet television. In Jay G. Blumler, Jack M. McLeod & Karl

Erik Rosengren (eds.), *Comparatively speaking: Communication and culture across space and time* (pp. 121-139). London: Sage.
Hallin, Daniel C. & Mancini, Paolo (2004). *Comparing media systems: Three models of media and politics*. Cambridge: University Press.
Hahn, Julia, Mok, Kathrin, Roessler, Patrick, Schmid, Michaela & Schwendemann, Nicolas (2008). Mediated events in political communication: A case study on the German European Council presidency 2007. *Communications, 33*(3), pp. 331-350.
Hanitzsch, Thomas, Hanusch, Folker, Mellado, Claudia, Anikina, Maria, Berganza, Rosa, Cangoz, Incilay, Coman, Mihai, Hamada, Basyouni, Hernandez, Maria Elena, Karadjov, Christopher D., Moreira, Sonia Virginia, Mwesige, Peter G., Plaisance, Patrick Lee, Reich, Zvi, Seethaler, Josef, Skewes, Elizabeth A., Vardiansyah Noor, Dani & Yuen, Kee Wang (2011). Mapping journalism cultures across nations: A comparative study of 18 countries. *Journalism Studies, 12*(3), pp. 273-293.
Held, David & McGrew, Anthony (2002). Introduction. In David Held & Anthony McGrew (eds.), *Governing globalization: Power, authority and global governance* (pp. 1-24). Cambridge: Polity Press.
Helfferich, Claudia (2005). *Die Qualität qualitativer Daten: Manual für die Durchführung qualitativer Interviews*. Wiesbaden: VS.
Hepp, Andreas (2006). *Transkulturelle Kommunikation*. Konstanz: UVK (UTB).
Hepp, Andreas (2009). Transculturality as a perspective: Researching media cultures comparatively. *Forum: Qualitative Research, 10*(1), article 26.
Hepp, Andreas, Bozdag, Cigdem & Suna, Laura (2011). *Mediale Migranten: Mediatisierung und die kommunikative Vernetzung der Diaspora*. Wiesbaden: VS.
Hepp, Andreas & Couldry, Nick (2010). Introduction: Media events in globalized media cultures. In Nick Couldry, Andreas Hepp & Friedrich Krotz (eds.), *Media events in a global age* (pp. 1-20). London: Routledge.
Hepp, Andreas & Krönert, Veronika (2005). Der katholische Weltjugendtag als Medienevent: Globalisierung der Medienkommunikation, deterritoriale religiöse Vergemeinschaftung und ‚branding religion'. In Stefan Nacke, René Unkelbach & Tobias Werron (eds.), *Weltereignisse: Theoretische und empirische Perspektiven* (pp. 151-175). Wiesbaden: VS Verlag.
Hepp, Andreas & Vogelgesang, Waldemar (2003). Einleitung: Ansätze einer Theorie populärer Events. In Andreas Hepp & Waldemar Vogelgesang (eds.), *Populäre Events: Medienevents, Spielevents und Spaßevents* (pp. 9-36). Opladen: Leske + Budrich.
Hepp, Andreas & Wessler, Hartmut (2009). Politische Diskurskulturen: Überlegungen zur empirischen Erklärung segmentierter europäischer Öffentlichkeit. *Medien & Kommunikationswissenschaft, 57*(2), pp. 174-197.
Hilgartner, Stephen & Bosk, Charles L. (1998). The rise and fall of social problems: A public arenas model. *American Journal of Sociology, 94*(1), pp. 53-78.
Hix, Simon (2005). *The political system of the European Union*. Basingstoke: Palgrave Macmillan.
Hunter, David (2010). Implications of the Copenhagen Accord for global climate governance. *Sustainable Development Law & Policy, 10*(2), pp. 4-15, 56-57.
Imber, Mark (1997). Geo-governance without democracy? Reforming the UN system. In Anthony McGrew (ed.), *The transformation of democracy?* (pp. 201-230). Cambridge: Polity Press.
Indymedia (2008, February 1). *Avaaz helped Japan climate shift*. Retrieved November 7, 2011 from http://de.indymedia.org/2008/01/206877.shtml.
Jahn, Detlef (2006). *Einführung in die vergleichende Politikwissenschaft*. Wiesbaden: VS.
Jamison, Andrew (1996). The shaping of the global environmental agenda: The role of non-governmental organisations. In Scott Lash, Bronislaw Szerszynski & Brian Wynne (eds.), *Risk, environment, and modernity* (pp. 224-245). London: Sage.

Johnson, Erik & McCarthy, John D. (2005). The sequencing of transnational and national social movement mobilization: The organizational mobilization of the global and U.S. environmental movements. In Donatella della Porta & Sidney Tarrow (eds.), *Transnational protest and global activism* (pp. 71-93). Lanham, MD: Rowman & Littlefield.

Kakonge, John O. (2011). The role of media in the climate change debate in developing countries. *Global Policy Essays*, November. Retrieved April 7, 2012 from http://www.globalpolicyjournal.com/sites/default/files/pdf/Kakonge%20Climate%20Change%20Essay.pdf.

Katz, Elihu & Liebes, Tamar (2007). No more peace! How disaster, terror and war have upstaged media events. *International Journal of Communication, 1*(1), pp. 157-166.

Keck, Margaret E. & Sikkink, Kathryn (1998a). *Activists beyond borders: Advocacy networks in international politics*. Ithaca, NY: Cornell University Press.

Keck, Margaret E. & Sikkink, Kathryn (1998b). Environmental advocacy networks. In Margaret E. Keck & Kathryn Sikkink (1998). *Activists beyond borders: Advocacy networks in international politics* (pp. 121-163). Ithaca, NY: Cornell University Press.

Khagram, Sanjeev, Riker, James V. & Sikkink, Kathryn (2002). From Santiago to Seattle: Transnational advocacy groups restructuring politics. In Sanjeev Khagram, James V. Riker & Kathryn Sikkink (eds.), *Restructuring world politics: Transnational social movements, networks, and norms* (pp. 3-23). Minneapolis: University of Minnesota Press.

King, Gary, Keohane, Robert O. & Verba, Sidney (1994). *Designing social inquiry: Scientific inference in qualitative research*. Princeton, NJ: Princeton University Press.

Kleinsteuber, Hans (2000). Öffentlichkeit und öffentlicher Raum. In Werner Faulstich & Knut Hickethier (eds.), *Öffentlichkeit im Wandel: Neue Beiträge zur Begriffsklärung* (pp. 34-48). Bardowick: Wissenschaftlicher Verlag.

Koopmans, Ruud & Statham, Paul (1999). Ethnic and civic conceptions of nationhood and the differential success of the extreme right in Germany and Italy. In Marco Giugni, Doug McAdam & Charles Tilly (eds.), *How social movements matter* (pp. 225-251). Minneapolis, MN: University of Minnesota Press.

Koopmans, Ruud & Statham, Paul (2010a). Theoretical framework, research design, and methods. In Ruud Koopmans & Paul Statham (eds.), *The making of a European public sphere: Media discourse and political contention* (pp. 34-59). Cambridge: University Press.

Koopmans, Ruud & Statham, Paul (eds.) (2010b). *The making of a European public sphere: Media discourse and political contention*. Cambridge: University Press.

Kozloff, Nikolas (2010, October 12). What is the Brazilian brand? *The Huffington Post*. Retrieved November 7, 2011 from http://www.huffingtonpost.com/nikolas-kozloff/what-is-the-brazilian-bra_b_759412.html.

Kriesi, Hanspeter (2004). Strategic political communication: Mobilizing public opinion in "audience democracies". In Frank Esser & Barbara Pfetsch (eds.), *Comparing political communication: Theories, cases, and challenges* (pp. 184-212). Cambridge: University Press.

Krippendorf, Klaus (2004). *Content analysis: An introduction to its methodology*. Thousand Oaks, CA: Sage.

Kuckartz, Udo (2010). *Einführung in die computergestützte Analyse qualitativer Daten*. Wiesbaden: VS.

Kuckartz, Udo, Dresing, Thorsten, Rädiker, Stefan & Stefer, Claus (2007). *Qualitative Evaluation: Der Einstieg in die Praxis*. Wiesbaden: VS.

Kunczik, Michael (2003). States, international organizations, and the news media: Problems of image cultivation. In Philippe J. Maarek and Gadi Wolfsfeld (eds.), *Political communication in a new era* (pp. 117-138). London: Routledge.

Kunelius, Risto & Nossek, Hillel (2008). Between the ritual and the rational: From media events to moments of global public spheres. In Elisabeth Eide, Risto Kunelius & Angela Phillips

(eds.), *Transnational media events: The Mohammed cartoons and the imagined clash of civilizations* (pp. 253-273). Göteborg: Nordicom.
Kyriakidou, Maria (2008). Rethinking media events in the context of a global public sphere: Exploring the audience of global disasters in Greece. *European Journal of Communication Research, 33*(3), pp. 273-292.
Latzer, Michael & Saurwein, Florian (2006). Europäisierung durch Medien: Ansätze und Erkenntnisse der Öffentlichkeitsforschung. In Langenbucher, Wolfgang & Latzer, Michael (eds.), *Europäische Öffentlichkeit und medialer Wandel* (pp. 10-44). Wiesbaden: VS.
L'Etang, Jacquie (2009). Public relations and diplomacy in a globalized world: An issue if public communication. *American Behavioral Scientist, 53*(4), pp. 607-626.
Liebes, Tamar (1998). Television disaster marathons: A danger of democratic processes? In Tamar Liebes & James Curran (eds.), *Media, ritual and identity* (pp. 71-84). London: Routledge.
Liebes, Tamar & Katz, Elihu (1997). Staging peace: Televised ceremonies of reconciliation. *Communication Review, 2*(2), pp. 235-257.
Lingenberg, Swantje (2009). The citizen audience and European transcultural public spheres: Exploring civic engagement in European political communication. *Communications, 35*(1), pp. 45-72.
Luhmann, Niklas (1990). Gesellschaftliche Komplexität und öffentliche Meinung. In Niklas Luhmann (ed.), *Soziologische Aufklärung 5: Konstruktivistische Perspektiven* (pp. 170-182). Opladen: Westdeutscher Verlag.
Lull, James (2007). *Culture-on-demand: Communication in a crisis world.* Malden MA: Blackwell Publishers.
Lynch, Marc (2005). *Voices of the new Arab public: Iraq, al-Jazeera, and Middle East politics today.* New York: Columbia University Press.
Marcinkowski, Frank (1993). *Publizistik als autopoietisches System – Politik und Massenmedien: eine systemtheoretische Analyse.* Opladen: Westdeutscher Verlag.
Mayring, Philipp (1983). *Qualitative Inhaltsanalyse: Grundlagen und Techniken.* Weinheim: Deutscher Studienverlag.
Mayring, Philipp (2003). *Qualitative Inhaltsanalyse: Grundlagen und Techniken.* Weinheim: Beltz.
McAdam, Doug (1996). Conceptual origins, current problems, future directions. In Dough McAdam, John D. McCarthy & Mayer N. Zald (eds.), *Comparative perspectives on social movements: Political opportunities, mobilizing structures, and cultural framings* (pp. 23-40). Cambridge: University Press.
McCurdy, Patrick M. (2008). Inside the media event: Examining the media practices of Dissent! at the Hori-Zone eco-village at the 2005 G8 Gleneagles summit. *Communications, 33*(3), pp. 295-311.
McGrew, Anthony (1997). Globalization and territorial democracy: An introduction. In Anthony McGrew (ed.), *The transformation of democracy?* (pp. 1-24). Cambridge: Polity Press.
McQuail, Denis (2010). *McQuail's mass communication theory.* London: Sage.
Melissen, Jan (2003). Summit diplomacy coming of age. *Discussion Papers in Diplomacy* of the Netherlands Institute of International Relations 'Clingendael', 86. Retrieved May 1, 2012 from http://www.clingendael.nl/publications/2003/20030500_cli_paper_dip_ issue86.pdf.
Melissen, Jan (2007). The new public diplomacy: Between theory and practice. In Jan Melissen (ed.), *The new public diplomacy: Soft power in international relations* (pp. 3-27). Basingstoke: Palgrave Macmillan.
Merten, Klaus (1992). Begriff und Funktion der Public Relations. *PR-Magazin, 23*(11), pp. 35-46.
Midttun, Atle, Coulter, Paddy, Gadzepko, Audrey, Wang, Jin, and Staurem, Elin (2011). Climate policy shifts across three continents: Media coverage in Norway, China and Ghana. *CERES 21 Media Reports*, 5. Retrieved April 7, 2012 from http://www.ceres21. org/media/User Media/20111125%20CERES21%20Media%20report.pdf.

Miller, Norman. (2002). *Environmental politics: Interest groups, the media and the making of policy*. Washington, DC: Lewis.

Moon Palace (n.d.). *Moon Palace Golf & Spa Resort*. Retrieved November 7, 2011 from http://www.moon-palace.com.

Mueller, Sherry (2009). The nexus of U.S. public diplomacy and citizen diplomacy. In Nancy Snow & Philip M. Taylor (eds.), *Routledge handbook of public diplomacy* (pp. 101-107). New York: Routledge.

Nanz, Patrizia & Steffek, Jens (2004). Global governance, participation and the public sphere. *Government and Opposition*, *39*(2), pp. 314-335.

Neidhardt, Friedhelm (1994). Öffentlichkeit, öffentliche Meinung, soziale Bewegungen. In Friedhelm Neidhardt (ed.), *Öffentlichkeit, öffentliche Meinung, soziale Bewegungen* (pp. 7-41). Opladen: Westdeutscher Verlag.

Newell, Peter (2000). *Climate for change: Non-state actors and the global politics of the greenhouse*. Cambridge: University Press.

News Around Us (2008, June 18). *Climate in crisis, leaders asleep*. Retrieved November 7, 2011 from http://www.newsaroundus.com/2008/06/climate-in-crisis-leaders-asleep.html.

Nisbet, Matthew C. (2011). Clear vision for the next decade of public debate. *Climate Shift studies & reports*, summer. Retrieved April 7, 2012 from http://climateshiftproject.org/wp-content/uploads/2011/08/ClimateShift_report_June2011.pdf.

Noelle-Neumann, Elisabeth (1980). *Die Schweigespirale: Öffentliche Meinung – unsere soziale Haut*. München: Langen Müller.

Nye, Joseph S. (2008). Public diplomacy and soft power. *The ANNALS of the American Academy of Political and Social Science*, *616*, pp. 94-109.

O'Neil, Jim (2001). *Building better global economic BRICs* (Goldman Sachs global economics paper no. 66). Retrieved November 7, 2011 from http://www.goldmansachs.com/our-thinking/brics/building-better.html.

Oberthür, Sebastian, Buck, Matthias, Müller, Sebastian, Pfahl, Stephanie, Tarasofsky, Richard G., Werksmann, Jacob & Palmer, Alice (2002). *Participation of non-governmental organisations in international environmental co-operation: Legal basis and practical experience*. Berlin: Erich Schmidt Verlag.

Olausson, Ulrika (2009). Global warming–global responsibility? Media frames of collective action and scientific certainty. *Public Understanding of Science*, *18*(4), pp. 421-436.

Olesen, Thomas (2005). Transnational publics: New spaces of social movement activism and the problem of global long-sightedness. *Current Sociology*, *53*(2), pp. 419-440.

Paul, Christopher (2011). *Strategic communication: Origins, concepts, and current debates*. Santa Barbara, CA: Praeger.

Padovani, Cinzia (2010). Citizens' communication and the 2009 G8 summit. *International Journal of Communication*, *4*, pp. 416-439.

Peters, Bernhard (2007). *Der Sinn von Öffentlichkeit (herausgegeben von Hartmut Wessler)*. Frankfurt am Main: Suhrkamp.

Peters, Bernhard, Sifft, Stefanie, Wimmel, Andreas, Brüggemann, Michael & Kleinen-von Königslow, Katharina (2005). National and transnational public spheres: the case of the EU. *European Review*, *13*(1), pp. 139-160.

Peters, Bernhard & Wessler, Hartmut (2006). Transnationale Öffentlichkeiten: Analytische Dimensionen, normative Standards, sozialkulturelle Produktionsstrukturen. In Kurt Imhof, Roger Blum, Heinz Bonfadelli & Otfried Jarren (eds.), *Demokratie in der Mediengesellschaft* (pp. 125-144). Wiesbaden: VS.

Pfetsch, Barbara (2008). Agents of transnational debate across Europe: The press in emerging European public sphere. *Javnost – The Public*, *15*(4), pp. 21-40.

Pfetsch, Barbara & Wehmeier, Stefan (2002). Sprecher: Kommunikationsleistungen gesellschaftlicher Akteure. In Otfried Jarren & Hartmut Wessler (eds.), *Journalismus – Medien – Öffentlichkeit: Eine Einführung* (pp. 39-97). Wiesbaden: Westdeutscher Verlag.

Price, Monroe E. (1995). *Television, the public sphere and national identity*. Oxford: Clarendon Press.

Princen, Thomas (1994). NGOs: Creating a niche in environmental diplomacy. In Thomas Princen & Matthias Finger (eds.) (1994). *Environmental NGOs in world politics* (pp. 29-47). London: Routledge.

Princen, Thomas & Finger, Matthias (eds.) (1994). *Environmental NGOs in world politics*. London: Routledge.

Przyborski, Aglaja & Wohlrab-Sahr, Monika (2009). *Qualitative Sozialforschung: Eine Einführung*. München: Oldenbourg.

Putnam, Robert D. (1988). Diplomacy and domestic politics. *International Organization*, *42*(3), pp. 427-460.

Raeymaeckers, Karin, Cosijn, Lieven & Deprez, Annelore (2007). Reporting the European Union: An analysis of the Brussels press corps and the mechanisms influencing the news flow. *Journalism Practice*, *1*(1), pp. 102-119.

Raupp, Juliana (2011). Die Legitimation von Unternehmen in öffentlichen Diskursen. In Raupp, Juliana, Jarolimek, Stefan & Schultz, Friederike (eds.), *Handbuch CSR: Kommunikationswissenschaftliche Grundlagen, disziplinäre Zugänge und methodische Herausforderungen* (pp. 97-114). Wiesbaden: VS.

Raustiala, Kal (2001). Nonstate actors in the global climate change regime. In Urs Luterbacher & Detlef F. Sprinz (eds.), *International relations and global climate change* (pp. 95-117). Cambridge, MA: MIT Press.

Revkin, Andrew C. (2010, December 1). The ghost of Kyoto visits Cancún. *The New York Times* (online edition). Retrieved November 7, 2011 from http://dotearth.blogs.nytimes.com/2010/12/01/the-ghost-of-kyoto-visits-cancun.

Reynolds, David (2007). *Summits: Six meetings that shaped the twentieth century*. New York: Basic Books.

Riordan, Shaun (2004). Dialogue-based public diplomacy: A new foreign policy paradigm? *Discussion Papers in Diplomacy* of the Netherlands Institute of International Relations 'Clingendael', 95. Retrieved May 1, 2012 from http://www.clingendael.nl/publications/2004/20041100_cli_paper_dip_issue95.pdf.

Risse, Thomas (2000). "Let's argue!": Communicative action in world politics. *International Organization*, *54*(1), pp. 1-39.

Risse, Thomas & Sikkink, Kathryn (1999). The socialization of international human rights norms into domestic practices: Introduction. In Thomas Risse, Stephen C. Ropp, and Kathryn Sikkink (eds.), *The power of human rights: International norms and domestic change* (pp. 1-38). Cambridge: University Press.

Rivenburgh, Nany K. (2010). *In pursuit of a global image: Media events as political communication*. In Nick Couldry, Andreas Hepp & Friedrich Krotz (eds.), *Media events in a global age* (pp. 187-202). London: Routledge.

Röttger, Ulrike, Preusse, Joachim & Schmitt, Jana (2011). *Grundlagen der Public Relations: Eine kommunikationswissenschaftliche Einführung*. Wiesbaden: VS.

Rosenau, James N. (1995). Governance in the twenty-first century. *Global Governance*, *1*(1), pp. 13-43.

Rucht, Dieter (1999). The impact of environmental movements in Western societies. In Marco Giugni, Doug McAdam & Charles Tilly (eds.), *How social movements matter* (pp. 204-224). Minneapolis, MN: University of Minnesota Press.

Rucht, Dieter (2001). Transnationaler politischer Protest im historischen Längsschnitt. In Ansgar Klein, Ruud Koopmans & Heiko Geiling (eds.), *Globalisierung – Partizipation – Protest* (pp. 77-96). Opladen: Leske + Budrich.

Russel, Christine (2010). From Copenhagen to Cancun: A challenging year for the climate story. *Columbia Journalism Review*, November 24. Retrieved April 7, 2012 from http://www.cjr.org/the_observatory/from_copenhagen_to_cancun.php.

Price, Monroe E. (2009). The end of television and foreign policy. *The ANNALS of the American Academy of Political and Social Science*, *625*, pp. 196-204.

Sarcinelli, Ulrich (2005). *Politische Kommunikation in Deutschland: Zur Politikvermittlung im demokratischen System.* Wiesbaden: VS.

Schäfer, Mike S., Ivanova, Ana & Schmidt, Andreas (2011). Globaler Klimawandel, globale Öffentlichkeit? Medienaufmerksamkeit für den Klimawandel in 23 Ländern. *Studies in Communication | Media, 1*, pp. 131-148.

Scharpf, Fritz W. (1998). Demokratie in der transnationalen Politik. In Wolfgang Streeck (ed.), *Internationale Wirtschaft, nationale Demokratie: Herausforderungen für die Demokratietheorie* (pp. 151-174). Frankfurt am Main: Campus.

Schill, Dan (2009). *Stagecraft and statecraft: Advance and media events in political communication.* Lanham, MD: Lexington Books.

Schmidt, Andreas (2012). Bewegungen, Gegenbewegungen, NGOs: Klimakommunikation zivilgesellschaftlicher Akteure. In Irene Neverla & Mike S. Schäfer (eds.), *Das Medien-Klima: Fragen und Befunde der kommunikationswissenschaftlichen Klimaforschung* (pp. 69-94). Wiesbaden: Springer VS.

Schmidt, Manfred G. (2010). *Demokratietheorien: Eine Einführung.* Wiesbaden: VS.

Schmitt-Beck (2001). Soziale Bewegungen und Öffentlichkeit: Theoretische Anmerkungen zu Bedeutung, Formen und Problemen der Öffentlichkeitsarbeit nicht-etablierter Kollektivakteure. In Günter Bentele, Tobias Liebert & Michael Vogt (eds.), *PR für Verbände und Organisationen: Fallbeispiele aus der Praxis* (pp. 15-35). Neuwied: Luchterhand.

Schwägerl, Christian & Traufetter, Gerald (2010, December 21). Germany grows tired of leading Europe on climate change. *Spiegel Online International.* Retrieved November 7, 2011 from http://www.spiegel.de/international/germany/0,1518,735594,00.html.

Seitz, John L. & Hite, Kristen A. (2012). *Global issues: An introduction.* Malden, MA: Wiley-Blackwell.

Seyfang, Gil (2003). Environmental mega-conferences – from Stockholm to Johannesburg and beyond. *Global Environmental Change*, *13*(3), pp. 223-228.

Shanahan, Mike (2009). Time to adapt? Media coverage of climate change in nonindustrialised countries. In Tammy Boyce & Justin Lewis (eds.), *Climate change and the media* (pp. 145-157). New York: Peter Lang Publishing.

Sharp, Paul (2007). Revolutionary states, outlaw regimes and the techniques of public diplomacy. In Jan Melissen (ed.), *The new public diplomacy: Soft power in international relations* (pp. 106-123). Basingstoke: Palgrave Macmillan.

Sheafer, Tamir & Gabay, Itay (2009). Mediated public diplomacy: A strategic contest over international agenda building and frame building. *Political Communication*, *26*(4), pp. 447-467.

Signitzer, Benno H. & Coombs, Timothy (1992). Public relations and public diplomacy: Conceptual convergences. *Public Relations Review*, *18*(2), pp. 137-147.

Sikkink, Kathryn (2002). Restructuring world politics: The limits and asymmetries of soft power. In Sanjeev Khagram, James V. Riker & Kathryn Sikkink (eds.), *Restructuring world politics: Transnational social movements, networks, and norms* (pp. 301-317). Minneapolis: University of Minnesota Press.

Sinclair, John (2005). International television channels in the Latin American audiovisual space. In Jean K. Chalaby (ed.), *Transnational television channels worldwide: Towards a new media order* (pp. 196-215). London: I. B. Tauris

Sireau, Nicolas (2009). *Make poverty history: Political communication in action*. Basingstoke: Palgrave Macmillan.

Snow, Nancy (2009). Rethinking public diplomacy. In Nancy Snow & Philip M. Taylor (eds.), *Routledge handbook of public diplomacy* (pp. 3-11). New York: Routledge.

Snow, Nancy & Taylor, Philip M. (2009). Preface and introduction. In Nancy Snow & Philip M. Taylor (eds.), *Routledge handbook of public diplomacy* (pp. ix-xi). New York: Routledge.

Sparks, Collin (2005). Media and the global public sphere: An evaluative approach. In Wilma de Jong, Martin Shaw & Stammers, Neil (eds.), *Global activism, global media* (pp. 34-49). London: Pluto Press.

Stammers, Neil & Eschle, Catherine (2005). Social movements and global activism. In Wilma de Jong, Martin Shaw & Stammers, Neil (eds.), *Global activism, global media* (pp. 50-67). London: Pluto Press.

Stevie Awards (2010, August 26). *Green diplomacy: Brazilian Secretariat of Social Communication*. Retrieved November 7, 2011 from http://thestevies.com/StevieAwardsUpdate/August26201 0fulltext.html.

Szyska, Peter (2009). Die Leistung der PR-Arbeit in der Marken- und Produktkommunikation. In Nina Janich (ed.), *Marke und Gesellschaft: Markenkommunikation im Spannungsfeld von Werbung und Public Relations* (pp. 17-52). Wiesbaden: VS.

Take, Ingo (2007). Zwischen Lobbyismus und Aktivismus: Der Wandel der Einflussstrategien von Umweltverbänden auf internationaler Ebene. In Ralf Kleinfeld, Annette Zimmer & Ulrich Willems (eds.), *Lobbying in Deutschland, den USA und in der Europäischen Union: Strukturen – Akteure – Strategien* (pp. 196-216). Wiesbaden: VS.

TckTckTck (2010, December 9). *Movie-themed advert asks Japan: Is a climate treaty 'Washed Away'?* Retrieved November 7, 2011 from http://tcktcktck.org/2010/12/movie-themed-advert-asks-japan-is-a-climate-treaty-washed-away/.

Teriete, Christian (2011). *Modern talking: Communicating the race to the future*. Presentation given in the seminar 'Communicating climate change globally' at the University of Mannheim on October 26.

The Japan Times (2010, November 27). *Japan will oppose Kyoto extension at COP16*. Retrieved November 7, 2011 from http://www.japantimes.co.jp/text/nn20101127a4.html.

Thörn, Håkan (2007). Social movements, the media and the emergence of a global public sphere: From anti-apartheid to global justice. *Current Sociology, 55*(6), pp. 896-918.

Thussu, Daya K. (2007). Mapping global media flow and contra-flow. In Daya K. Thussu (ed.), *Media on the move: Global flow and contra-flow* (pp. 11-32). London: Routledge.

Tobler, Stefan (2006). Konfliktinduzierte Transnationalisierung nationaler und supranationaler Öffentlichkeitsarenen: Indikatoren einer europäischen Öffentlichkeit. In Langenbucher, Wolfgang & Latzer, Michael (eds.), *Europäische Öffentlichkeit und medialer Wandel* (pp. 105-130). Wiesbaden: VS.

USC Center on Public Diplomacy (2009). *What is public diplomacy?* Retrieved June 8, 2009 from http://uscpublicdiplomacy.org/index.php/about/what_is_pd/.

UNFCCC (2010). *Standard admission process for non-governmental organizations (NGOs)*. Retrieved June 27, 2012 from http://unfccc.int/files/parties_and_observers/ngo/application/pdf/admission_process_2010_english.pdf.

UNFCCC (n.d.). *Participation breakdown*. Retrieved November 7, 2011 from http://unfccc.int/files/parties_and_observers/ngo/application/pdf/participation_breakdown_cop_1-16.pdf.

Van Audenhove, Leo (2007). *Expert interviews and interview techniques for policy analysis*. Presentation at Vrije Universiteit Brussel retrieved June 8, 2009 from http://www.ies.be/files/060313%20Interviews_VanAudenhove.pdf.

Van Ham, Peter (2007). Power, public diplomacy, and the *Pax Americana*. In Jan Melissen (ed.), *The new public diplomacy: Soft power in international relations* (pp. 47-66). Basingstoke: Palgrave Macmillan.

Vidal, John (2010a, December 1). Cancún climate change summit: Japan refuses to extend Kyoto protocol. *The Guardian* (online edition). Retrieved November 7, 2011 from http://www.guardian.co.uk/environment/2010/dec/01/cancun-climate-change-summit-japan-kyoto.

Vidal, John (2010b, December 8). Cancún climate change summit: Japan defiant in face of diplomatic pressure. *The Guardian* (online edition). Retrieved November 7, 2011 from http://www.guardian.co.uk/environment/2010/dec/08/cancun-climate-change-summit-japan.

Visible Measures (2012, March 9). *Kony social video campaign fastest growing in history*. Retrieved April 7, 2012 from http://corp.visiblemeasures.com/news-and-events/blog/bid/79508/Kony-Social-Video-Campaign-Fastest-Growing-in-History.

Vlahos, Michael (2009) Public diplomacy as loss of world authority. In Nancy Snow & Philip M. Taylor (eds.), *Routledge handbook of public diplomacy* (pp. 24-38). New York: Routledge.

Volkmer, Ingrid (1999a). International communication theory in transition: Parameters of the new global public sphere. *MIT Communications Forum*. Retrieved April 7, 2012 from http://web.mit.edu/comm-forum/papers/volkmer.html.

Volkmer, Ingrid (1999b). *News in the global sphere: A study of CNN and its impact on global communication*. Luton, UK: University of Luton Press.

Volkmer, Ingrid (2003). The global network society and the global public sphere. *Development, 46*(1), pp. 9-16.

Voss, Kathrin (2007). *Öffentlichkeitsarbeit von Nichtregierungsorganisationen: Mittel – Ziele – interne Strukturen*. Wiesbaden: VS.

Vowe, Gerhard (2006). Feldzüge um die öffentliche Meinung: Politische Kommunikation in Kampagnen am Beispiel von Brent Spar und Muroroa. In Ulrike Röttger (ed.), *PR-Kampagnen: Über die Inszenierung von Öffentlichkeit* (pp. 125-150). Wiesbaden: VS.

Waltz, Kenneth (1979). *Theory of international politics*. Boston, MA: McGraw-Hill.

Warkentin, Craig (2001). *Reshaping world politics: NGOs, the Internet, and Global Civil Society*. Lanham, MD: Rowman & Littlefield.

Weilemann, Peter R. (2000). The summit meeting: The role and agenda of diplomacy at its highest level. *NIRA Review*, *7*(2), pp. 16-20

Weingart, Peter, Engels, Anita & Pansegrau, Petra (2000). Risks of communication: Discourses on climate change in science, politics, and the mass media. *Public Understanding of Science*, *9*(3), pp. 261-283.

Weiss, Thomas G. (2008). *What's wrong with the United Nations and how to fix it*. Cambridge: Polity Press.

Wesel, Reinhard (2004). *Symbolische Politik der Vereinten Nationen: Die „Weltkonferenzen" als Rituale*. Opladen: Leske + Budrich.

Wessler, Hartmut (2009, January). *Die Bedeutung von Medienevents für die Entstehung transnationaler Öffentlichkeit: Überlegungen für zukünftige Forschung*. Presentation at the Research Colloquium "Medienkultur" [="Media Culture"] at the Institute of Media, Communication, and Information Research, University of Bremen, Germany.

Wessler, Hartmut (2010). Public sphere. In *Oxford Bibliographies Online*, subject communication (http://www.oxfordbibliographiesonline.com). Oxford: University Press.

Wessler, Hartmut (2012). Identifying global public sphere moments. In Jan Fredrik Hovden and Karl Knapskog (eds.), *Hunting high and low: Skriftfest til Jostein Gripsrud på 60-årsdagen* (pp. 437-455). Oslo: Scandinavian Academic Press.

Wessler, Hartmut & Adolphsen, Manuel (2011, October). *Nachhaltige Medienevents? Strategische, rituelle und diskursive Aspekte langfristiger transnationaler Mediendebatten.* Presentation at the Three-Nation Conference "The New Structural Transformation of the Public Sphere", University of Innsbruck, Austria.

Wessler, Hartmut & Brüggemann, Michael (2012). *Transnationale Kommunikation: Eine Einführung.* Wiesbaden: Springer VS.

Wessler, Hartmut, Peters, Bernhard, Brüggemann, Michael, Kleinen-von Königslow, Katharina & Sifft, Stefanie (2008). *Transnationalization of public spheres.* Basingstoke: Palgrave Macmillan.

Wikipedia (n.d.). *Spirited Away.* Retrieved November 7, 2011 from http://en.wikipedia.org/wiki/Spirited_Away.

Wilcox, Dennis L., Ault, Phillip H. & Agee, Warren K. (1992). *Public relations: Strategies and tactics.* New York: HarperCollins.

Wimmer, Jeffrey (2007). *(Gegen-)Öffentlichkeit in der Mediengesellschaft: Analyse eines medialen Spannungsverhältnisses.* Wiesbaden: VS.

Yearley, Steven (1995). The transnational politics of the environment. In James Anderson, Chris Brook, and Allan Cochrane (eds.), *A global world? Re-ordering political space* (pp. 209-247). Oxford: University Press.

Yearley, Steven (2003). Social movements as problematic agents of global environmental change. In Steven Vertovec & Darrell Posey (eds.), *Globalization, globalism, environments, and environmentalism: The Linacre Lectures 2000* (pp. 39-54). Oxford: University Press.

Zehr, Stephen C. (2000). Public representations of scientific uncertainty about global climate change. *Public Understanding of Science*, *9*(2), pp. 85–103.

MIX
Papier aus verantwortungsvollen Quellen
Paper from responsible sources
FSC® C105338

If you have any concerns about our products,
you can contact us on
ProductSafety@springernature.com

In case Publisher is established outside the EU,
the EU authorized representative is:
Springer Nature Customer Service Center GmbH
Europaplatz 3, 69115 Heidelberg, Germany

Printed by Libri Plureos GmbH
in Hamburg, Germany